Electric Telegraph Series

THE ELECTRIC TELEGRAPH
ITS
HISTORY AND PROGRESS

NUMBER 2 IN THE ELECTRIC TELEGRAPH SERIES

COVER PICTURE
Lady telegraphists sending and receiving messages on
Cooke and Wheatstone needle telegraphs.

Edward Highton

The Electric Telegraph
its
History and Progress

TGR Renascent Books
2017

PUBLICATION HISTORY

THE ELECTRIC TELEGRAPH
ITS HISTORY AND PROGRESS

EDWARD HIGHTON

First Edition
1852

This new edited edition published by
TGR Renascent Books
27 Springdale Court
Mickleover, Derby DE3 9SW
United Kingdom
2017

www.renascentbooks.co.uk

All rights reserved.
No part of this publication may be reproduced,
stored in a retrieval system, or transmitted,
in any form or by any means, without the
prior permission in writing of the
publisher, or as expressly
permitted by law.

© Publisher's Copyright: TGR Renascent Books 2017

ISBN 978-1-9791199-9-3

CONTENTS

Chapter   *Page*

    List of Figures iv
    Publisher's Note vi
    Series Editor's Forward vii
    Author's Preface 1

1   Introduction.. 5

2   Telegraphs Generally... 7
      Telegraphs by Means of Light 7
      Telegraphs by Direct Sight 8
      Telegraphs by Electricity 10

3   On the Production of Electricity 13
      Electricity from Friction 13
      Electricity from Chemical Action 16
      On the Production of Electricity from the Magnet.......... 20
      Production of Electricity by a Change of Temperature 22
      Electricity from Animals...23
      Electricity from Other Sources 26

4   Methods for Detecting Electrical Signals 29
      On the Means Usually Employed for Transmitting Electricity
      to a Distant Place 34

5   First Principles of an Electric Telegraph 39

6   Telegraphs Prior to 1837.. 43
      Brief Summary of Telegraphs Prior to the Year 1837..........43
      Description of Telegraphs Prior to 1837 45
          Lesarge's Telegraph 45
          Lomond's Telegraph 46
          Betancourt's Telegraph.. 46

Contents

Chapter		Page
	Reizen's Telegraph	47
	Cavallo's Telegraph	47
	Salva's Telegraph	49
	Soemmering's Telegraph	49
	Schwieger's Telegraph	54
	Coxe's Telegraph	54
	Ronalds' Telegraph	55
	Ampére's Telegraph	58
	Triboaillet's Telegraph	60
	Schilling's Telegraph	61
	Gauss and Weber's Telegraph	61
	Taquin and Ettieyhausen Telegraph	63
	Steinheil's Telegraph	63
	Masson's Telegraph	64
	Morse's Telegraph	64
	Vail's Telegraph	67
	Alexander's Telegraph	68
	Davy's Telegraph;	69
	History of Telegraph Patents	70
7	**Telegraphs from 1837 to the Present Day**	**75**
	Cooke and Wheatstone's Telegraphs	75
	Cooke and Wheatstone's Five-Needle Telegraph	77
	Subsequent Patents by Cooke and Wheatstone	78
	Davy's Telegraph	86
	Bain's Telegraphs	88
	H & E Highton's Telegraphs	91
	Highton's Telegraphs	93
	Nott's Telegraph	104
	Pool's Patent	106
	Brett & Little's Telegraph	107
	Henley & Foster's Telegraph	107
8	**Current Transmission & Atmospheric Disturbances**	**111**
	Law of the Transmission of the Electric Current	111
	Atmospheric Disturbances of the Electric Telegraph	112
9	**Requirements for an Electric Telegraph**	**117**
	Keys	119
	Alarums	121
	Description of Alarum No. 1	122
	Description of Alarum No. 2	123
	Indicating Instruments	125

Contents

Chapter		Page
	Insulation	128
	Coils	130
	Quantity of Electricity to be Used for Telegraphing	130
	Lightning Conductors	131
	Subterranean Wires as Compared with Wires in the Air	132
	On the Velocity of the Electric Current	132
10	**Electric Telegraphs now in Use**	**137**
	Invention of the Electric Telegraphs now in Use	137
	Prevailing Ignorance as to the Action of the Electric Telegraph	138
	Telegraphs in Great Britain	140
	Extent of Electric Telegraphs in use in Great Britain	142
	Telegraphs in America	143
	Extent of Electric Telegraphs in America	144
	Telegraphs in France	146
	Extent of Telegraphs in France	147
	Telegraphs in Prussia and Germany	147
	Telegraphs in Russia and Other Countries	148
	Telegraphs in India	149
	On the Restrictions Imposed on the Use of the Electric Telegraph	151
	Telegraphs in the Bank of England	152
	Electric Telegraphs in America in Connection with the Fire Establishments	153
11	**Submarine Telegraphs**	**155**
	Telegraphing without Insulation	158
12	**Message Charges and the Regulation of Time**	**161**
	Charges for the Use of the Electric Telegraph	161
	Telegraph Charges in France and Prussia	163
	Regulation of Time by the Telegraph	168
	Electric Clocks	168
	Electric Clocks at the Observatory at Greenwich	169
13	**Utility of the Electric Telegraph**	**173**
	Concluding Remarks	177

iii

LIST OF FIGURES

Figure		Page
1	Facsimile of the first edition 1852 title page	xiv
2	Electric circuit between London and Liverpool	39
3	Galvanic battery in circuit between London and Liverpool	39
4	Needle telegraph circuit between London and Liverpool	40
5	Reizen telegraph tinfoil matrix, with breaks that shape letters and numerals	48
6	Soemmerings's telegraph, which used the decomposition of water by electricity to designate letters and numerals	51
7	Ronald's pith-ball and screen telegraph and electricity generator	56
8	Enlargement of Ronald's moveable dial and screen	56
9	Steinheil's telegraph	62
10	Transmitting apparatus of Steinheil's telegraph	62
11	Coils and magnets of Steinheil's telegraph	62
12	Morse key for sending currents of electricity	64
13	Morse telegraph recording instrument	65
14	Morse American alphabet	65
15	Alexander's indicating telegraph	69
16	Cooke & Wheatstone's five-needle telegraph	77
17	Cooke & Wheatstone double needle instrument (exterior view)	80
18	Cooke & Wheatstone double needle instrument (interior arrangement)	80
19	Cooke & Wheatstone Electro-Magnetic telegraph	82
20	Intermediate telegraph post	84
21	Drawing post (side view)	84
22	Drawing post (front view)	84
23	Vertical section of Figs. 21 & 22	84
24	Davy's telegraph 1838 patent	87
25	Highton's system of marks	93

Figure		Page
26	Highton gold-leaf telegraph	94
27	Highton gold-leaf signalling apparatus	95
28	Highton patented telegraph arrangements	97
29	Highton single-pointer telegraph	98
30	Highton double-pointer telegraph	98
31	Highton revolving-pointer telegraph	99
32	Highton revolving-disc telegraph	100
33	Highton direct letter-showing telegraph	100
34	Highton printing telegraph mechanism	101
35	Highton printing telegraph mechanism—one line wire	101
36	Highton printing telegraph mechanism—multiple line wires	101
37	Nott telegraph instrument	105
38	Nott telegraph (interior view)	105
39	Henley & Forster magnetic telegraph instrument	108
40	Henley & Forster magnetic telegraph (interior view)	108
41	Magnetic needle and electro-magnet	108
42	Lightning damage at Oundle Station	114
43	Lightning damage at Thrapston Station (front view)	114
44	Lightning damage at Thrapston Station (back view)	114
45	Highton telegraph keys (side view)	120
46	Highton telegraph keys (top view)	120
47	Highton's alarum No. 1 (view inside case)	122
48	Highton's alarum No. 1 (side view)	122
49	Highton's alarum No. 2 (view inside case)	124
50	Highton's alarum No. 2 (top view)	124
51	Highton disc telegraph	126
52	Highton disc telegraph (internal features)	126
53	Experiment to determine the speed of electricity	134
54	Alphabet code used by The Electric Telegraph Company	141
55	Mr Jacob Brett's telegraph	143
56	Breguet's telegraph as used in France	145
57	Breguet's handle apparatus	145
58	Alphabetical code used with Breguet's telegraph	146
59	Prussian adaptation of Morse	147
60	Submarine cable laid in the English Channel	156

PUBLISHER'S NOTE

The international telegram service in Britain, inaugurated by private enterprise in 1845 but soon taken over by the General Post Office (GPO), and latterly by British Telecom (BT), ended in 2003. In the United States the service finished when Western Union sent its last telegram in 2006. As a consequence, most people today, in the age of the Internet, Satellite Communications, Mobile Phones, E-mail, Instant Text Messaging and Fibre-Optic Cable, have no idea that the world was once girdled with thin iron wires strung on poles over thousands of miles of often inhospitable terrain, or that armoured cables lay fathoms deep in abysmal darkness on the bottom of the oceans. It was via these fragile threads that the world once communicated.

The purpose of this Electric Telegraph Series is to publish in new editions some of the many books on telegraphy that first appeared in the Victorian era. Neglected and forgotten, dismissed as no longer relevant, these books are a treasure trove for historians of technology, research students and interested lay persons. The technology and operation of the telegraph very quickly achieved a level of development and sophistication that now seems quite staggering, as a perusal of the books in this series will soon show.

One caveat must be mentioned—the men writing these books were in complete ignorance of the *nature* of electricity, although of course fully conversant with its *effects*. Electricity was often called a "mysterious fluid," by Victorians on the analogy that electricity somehow flowed through a wire like water flows through a pipe. It was not until 1897 that the atom was "split" and J. J. Thomson discovered the electron, the sub-atomic particle that is ultimately responsible for the flow of electricity. It was well into the twentieth century before a coherent theory of electricity was developed and promulgated. Be cautious, therefore, when reading early authors on electricity. Those who have the need should consult professionals or up-to-date text-books on the subject. For the remainder, the books in this series will provide wonderfully readable and easy to understand accounts of electricity, which while not always strictly accurate, nevertheless provide everything needed to understand telegraphy, telegraphic circuits, telegraphic instruments and their ubiquitous power sources—hand-turned generators or wet batteries of exceptional size.

SERIES EDITOR'S FORWARD

The author of this book, Edward Highton, was born on the 13th August, 1817, at Leicester, England. He was the youngest son of Henry Highton of that town. His elder brother Henry, named after his father, was born in the previous year, on 19th January, 1816. Although quite different in temperament and in the careers they followed, the brothers remained close throughout their lives. Both were to play an important part in the development of the electric telegraph.

As the first born, it was natural in that age that Henry should be favoured by his parents. He was educated at Rugby School under the famous headmaster Thomas Arnold, with whom he remained friendly for many years afterwards. On 13th March, 1834 he matriculated at Queens College, Oxford, and in 1837 graduated BA with a first class in classics. He proceeded MA in 1840 and was Michel fellow of his college in 1840-41. In the same year he began tutoring the mathematician Henry Smith.

Edward was denied the prestigious education enjoyed by his brother, perhaps because of a lack of money in the family. Nevertheless, his education was not neglected and he was sent first to the grammar school in Leicester, and then latterly to a private tutor, the Reverend John Foster. By these means he received a mathematical and classical education that seems to have been in no way inferior to that of his brother. He then left home to become the pupil of Mr Stephen Robinson, (M. Inst. C.E.), of Durham, the engineer to the Hartlepool Dock and Railway Company. Meanwhile, Henry had married Elizabeth Paxton, and on his marriage was appointed Assistant Master at Rugby School. He then progressed to an appointment as Principal at Cheltenham College. While there he published a revised translation of the New Testament, several sermons and some notable theological works. By this time he was an ordained minister in the Anglican Church. Edward never married, but throughout his life he supported his three sisters and became, in time, a much loved uncle to Henry and Elizabeth's twelve children.

After completing his professional education, Edward was entrusted by Robinson with the management of the works at Hartlepool, after the contactors failed to perform the excavations required. Promoted to Assistant Engineer, he then superintended for upwards of three years the construction of the docks on behalf of the company. As a result of this experience, in 1845 he was appointed Resident Engineer of the Taff Vale Dock and Railway Company, in South Wales. In the meantime, in the true spirit of the Victorian amateur gentleman, Henry was amusing himself by conducting a number of practical experiments in the application of electricity to telegraphy. In the mid-1840s he purchased the rights to exclusive use of a gold-leaf instrument, which he proceeded to develop and adapt for telegraph purposes.

Edward became very interested in Henry's experiments, and thus began their collaboration in the development of telegraphic apparatus, which lasted from the 1840s until the 1870s. So great was Edward's enthusiasm that he devoted himself to the especial study of electric telegraphy, with the result that in 1846 he became Telegraphic Engineer to the London & North Western Railway. During this time he contributed the whole of the mechanical details and many of the principles of the several patents taken out by himself and Henry. In 1847 he became a member of the Institution of Civil Engineers, and took part in the discussions upon telegraphic subjects at the meetings of that institute. He read several papers before the Society of Arts and one of them, "On Atmospheric Electricity", was published in the society's Transactions. In 1849 he received the large gold medal of the Society for his inventions in electric telegraphy. Subsequently, he received another medal at the Great Exhibition in 1851.

Henry had patented a high-tension telegraph in 1844, and the very sensitive gold-leaf telegraph in 1846. Edward was also busy developing a simplified and inexpensive needle-telegraph,

Series Editor's Forward

which utilised tappers, or keys, rather than the commutators with drop-handles usual in this form of instrument. He patented this apparatus in 1848. After his needle-telegraph patent, he went on to make a number of innovations in overhead wire telegraphy. He was an early advocate of resin-insulated subterranean circuits. Perhaps not unexpectedly, Henry also took up the subject and by the early 1870s he believed the sensitivity of his gold-leaf instrument would enable it to allow transatlantic communication along underwater wires without insulation. This development, if successful, would greatly cut the cost of the undersea cable laying then being undertaken. This, said Henry, would allow "the poor emigrant to communicate with his family ... and put an end to the necessarily almost prohibitory rates which at present prevail". In 1872 he read a paper on the subject before the Society of Arts, which was well received, and the Society presented him with their silver medal. This did not quite match the prestigious gold medal which Edward had been awarded twenty-three years earlier.

Unfortunately, Henry's experiments had been largely conducted in fresh water and by 1873 he had to concede that such a system would not work in a saline medium. He therefore began campaigning for the adoption of the cheap method of insulation he had developed using vegetable tar and lead oxides. In improving his gold-leaf instrument for telegraphic purposes, Henry had also done research into galvanism and the improvement of batteries. He devised several new types of battery, one of which (a type of zinc carbon battery) found much favour among electroplaters.

From *circa* 1845 Edward lived in London, at Clarence Villa, 5 Gloucester Road, Regents Park. Here, from 1849 onwards, he founded and promoted the British Electric Telegraph Company, set up to work the Highton family patents. It was during this period that he found time to write this book, *The Electric Telegraph – its History and Progress*. In it he describes his and Henry's contributions to telegraphy, giving an account of Henry's gold-leaf telegraph among the other descriptions of Highton inventions. Perhaps because of declining health, which, it was said, had been "much injured by his professional labours and anxieties", he sold his interest in the British Electric Telegraph Company around 1855. Although compelled to retire from the active duties of his profession, his single-needle telegraph became one of the most widely used instruments in Britain.

Edward died on the 13th November, 1859, at the young age of 42 years. He was greatly beloved by his relatives, to many of whom he had almost acted the part of a father, and towards whom he had always shown the most kind and generous disposition. He predeceased his brother Henry, who spent his last years in Putney, London. Henry died at his home there, The Cedars, on 21st December, 1874, aged 58 years, his wife Elizabeth surviving him. The year before his death his battery patents were purchased and a company called the Highton Battery Company was set up to work them. It is probably true to say that the considerable contribution of the Highton brothers to electrical telegraphy remains to this day largely unrecognised.

Edward was the author of some poetry said to be of considerable merit, which was however, only printed for private distribution. This book on the electric telegraph was originally published in Weale's Series as an 8vo (octavo) edition, that is about 8 to 10 inches (200 to 250 mm) tall, with a width in proportion, in 1852. In it he reveals his classical education by numerous allusions to Greek authorities and by a few Latin quotations. His inclusion of a poem about the electric telegraph illustrates his love of poetry.

EDITOR'S COMMENTS

There are some references in this book that might puzzle a modern-day reader. One such is gutta percha, a name once familiar to every Victorian but now long out of memory. This substance became of tremendous importance in telegraphy as an insulator, especially of undersea cables. It is a rubbery latex resin obtained from the Palaquium fruit tree, found in the rain forests of the Malayan Archipelago. The Gutta Percha Company was formed early in the nineteenth century in London, at Wharf Road, Islington, to exploit the substance. Soon Victorians were being provided with gutta percha raincoats, boot soles, shoelaces, walking sticks, doorknobs, inkstands, snuff boxes and photograph frames, as well as brooches, earrings, lockets and buttons. Its use in golf balls trans-

Series Editor's Forward

formed the sport, but its true potential as a perfect insulator of submarine telegraph cables came when it was found to show no deterioration when submerged in salt water. It is true to say that few other materials have had such a revolutionary impact on the world, and few others have been so quickly forgotten.

Another is caoutchouc (pronounced cout-chuk), which is a natural rubber that has not been vulcanised. It is also known as India rubber, collected as latex resin by tapping rubber trees, mainly in India, Malaysia and Indonesia. Like gutta percha, it also was used for the insulation of telegraph cables but was considered an inferior alternative. Without vulcanisation, a process by which the rubber is heated and chemicals added to improve electrical resistance and to prevent it from perishing, it quickly deteriorated in service.

A word or two is perhaps appropriate to explain Highton's several references to the "old" Electric Telegraph Company. On 3rd September 1845 a syndicate of City merchants projected the first joint-stock company in the world, with the aim of uniting Britain through a network of electric communications. It was called "The Electric Telegraph Company", afterwards fondly referred to as "The Electric" and intended to exploit the telegraphic inventions of Charles Wheatstone and William Cooke. Highton seems to use the word "old" to distinguish this long-standing company from the later company promoted by the whole Highton family in 1850, which was named "The British Electric Telegraph Company". From the beginning, the company's managing director was Highton's nephew and namesake Edward, the son of Henry Highton. The "British Electric" was the first real challenger to the Electric company's patent monopoly, concentrating initially on Manchester, the principal centre for textile manufacture in the north-west of the country. This was an area the "old" company had not yet covered.

In the course of his treatise, Highton mentions many names of men engaged in the investigation of electricity and the invention of telegraphic apparatus. Men like Alessandro Volta, who gave his name to the Volt, the unit of electromotive force, and André-Marie Ampère, who gave his name to the Amp, the unit of electric current, or Michael Faraday, who contributed enormously to the study of electromagnetism and electrochemistry, are still well known today, at least to students of the history of science. Other names, like Professor Charles Wheatstone and William Fothergill Cooke are perhaps not so readily brought to mind, but are easily discoverable on the internet as important inventors and entrepreneurs in the field of electric telegraphy. Likewise the American Samuel Morse, whom most people today remember because of the everyday use, until very recent times, of his ubiquitous morse code.

However, many names quoted by Highton, all eminent and well known personages in his time, are today quite forgotten and have all but disappeared from the pages of history. When encountering such names, it is safe to assume that these people were Victorian amateurs, often clergymen with time on their hands, like Highton's elder brother, the Reverend Henry Highton. Others were those with inherited wealth, who, thus freed from the necessity of working or following a profession, were able to indulge their all consuming passions. The period was replete with amateurs, not only making important contributions to electric telegraphy, but adding to the sum of human knowledge by collecting and cataloguing plant, insect, fish and animal species, fossils, rocks, observing and recording atmospheric phenomena, the weather, the planets and the stars in the heavens. Forgotten most of them may be, but readers of this book will briefly engage with these pioneers, in the fields of electricity and telegraphy at least, through Highton's discussions.

Victorian fascination with just about everything in the natural world is manifest by Highton's discussion of "Electricity from Animals". Electric eels and other such creatures have absolutely nothing to do with electric telegraphy, but the phenomenon of electric current was such a new subject that it was deemed essential to present it in all its aspects. Highton was not alone in this, many other writers on telegraphy could not resist describing, with awe, the severe electric shocks received by those incautious enough to handle the creatures.

In a section entitled "History of Patents", Highton cannot resist the temptation to mount his high-horse. Obviously frustrated by

Series Editor's Forward

the law of patents as it existed in his day, he castigates the system and fulminates against its inflexibility. On reading his account, we cannot but sympathise with him. Two minor points remain to be mentioned. Highton often uses the word "alarum", which is simply an archaic word for 'alarm', as in the various alarm devices which alert telegraphists to the arrival of a transmitted signal. Finally, it should not be forgotten that when Highton refers to "the present day" he can only mean his "present", that is, the time at which he was writing his book, *circa* 1850.

Gordon Roberts
Derby, 2017

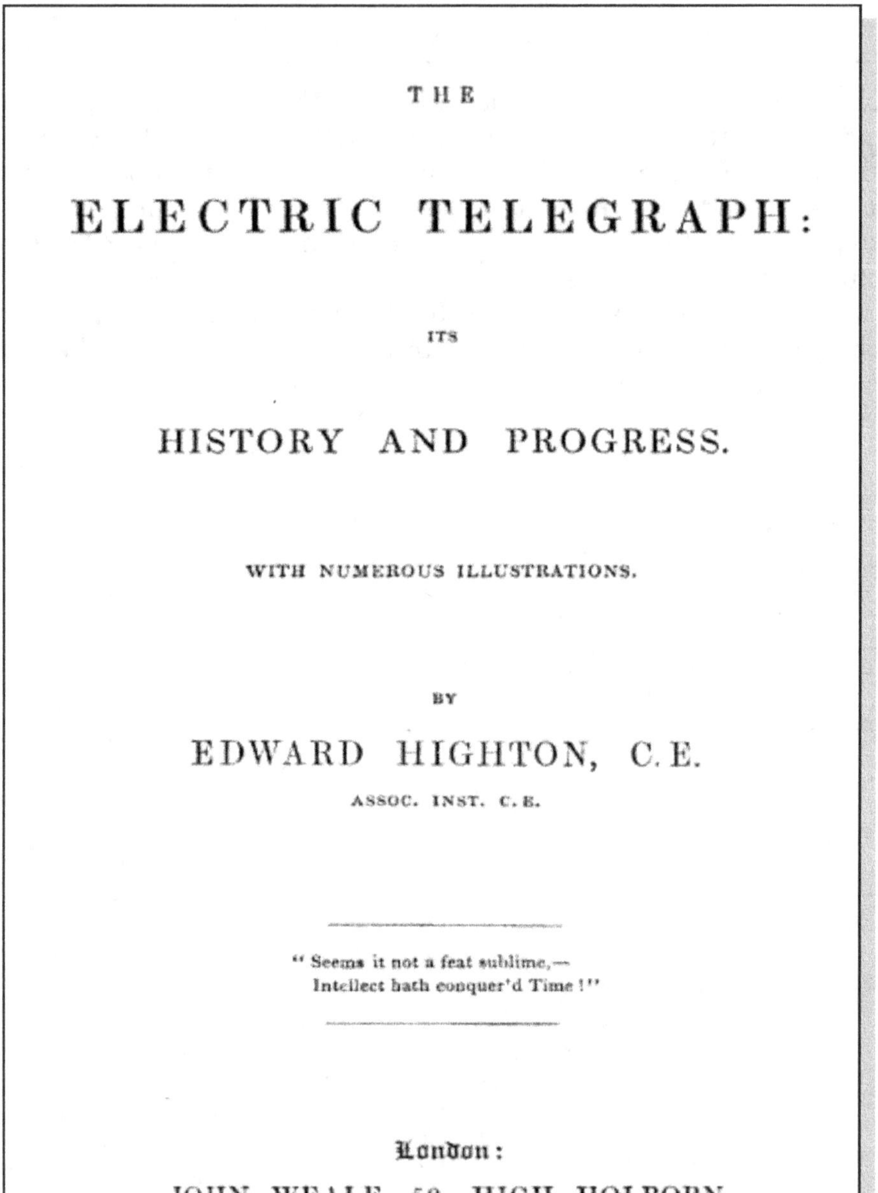

THE

ELECTRIC TELEGRAPH:

ITS

HISTORY AND PROGRESS.

WITH NUMEROUS ILLUSTRATIONS.

BY

EDWARD HIGHTON, C.E.

ASSOC. INST. C.E.

" Seems it not a feat sublime,—
Intellect hath conquer'd Time!"

London:
JOHN WEALE, 59, HIGH HOLBORN.
MDCCCLII.

Fig. 1
Facsimile of the first edition 1852 title page

AUTHOR'S PREFACE

It is the object of the following pages to give in as concise a manner as possible a description of the Electric Telegraph as a whole, and also to point out the peculiar features in several kinds of Telegraphs which are now in use. In a short treatise like this, to have described every plan in detail, of the many that have been proposed, would have been impossible; and it did not appear that a work so comprehensive would be at all suited to the general reader. In describing the various Telegraphs mentioned herein, every endeavour has been used to give a statement of the inventions, without venturing an opinion on the comparative merits of the plans, unless there was any peculiarity in the case which seemed to call for the expression of such opinion.

A succinct account of the Electric Telegraph has long been a desideratum in this kingdom: in foreign countries several works on the subject have already appeared. For many of the statements made herein the Author is indebted to these and other works, and also to the several scientific periodicals which are now so extensively published in this and other countries. As it would have been a source of considerable confusion to the general reader, for whom these pages have been written, to have been continually referring specifically to each work, the Author begs now freely and frankly to acknowledge the assistance which these publications have rendered to him, and the peculiar obligations under which he is placed with regard to their respective Authors.

The following is a list of some of the works which have been consulted, and from many of which copious extracts have been made.

- Annals of Electricity, Magnetism, and Chemistry, by W. Sturgeon.
- Alexander's Original Electric Telegraph.

- American Electro-Magnetic Telegraph, by A. Vail.
- Account of some remarkable Applications of the Electric Fluid to the useful Arts, by Alexander Bain.
- Book of the Telegraph, by Daniel Davis.
- Chamber's Papers for the People.
- Daniell's Introduction to Chemical Philosophy.
- Davis's Manual of Magnetism.
- Descriptions of an Electrical Telegraph, by F. Ronalds.
- Electrical Magazine.
- Encyclopaedia Britannica.
- Encyclopaedia Metropolitana.
- Engineer's and Mechanic's Pocket Book.
- Electric Telegraph Manipulation, by Charles V. Walker.
- Hand-Book of Chemistry, by Leopold Gmelin.
- Lardner's Cabinet Cyclopaedia (Electricity, by Dr. Lardner).
- London, Edinburgh and Dublin Philosophical Magazine.
- Noad's Electricity.
- Patent Journal.
- Practical Mechanic's and Engineer's Magazine.
- Repertory of Patent Inventions.
- Scientific Periodicals (various).
- Telegraphic Railways, by W. F. Cooke.
- Transactions of the Cambridge Philosophical Society.
- Transactions of the Society of Arts.
- Transactions of the Royal Society.
- Wonders of the Wire.
- Manuel de la Télégraphie Electrique, par L. Breguet.
- Rapport fait a l'Académie des Sciences sur les Appareils Télégraphiques. De M. Siemens (de Berlin).
- Traité Télégraphie Electrique, par M. L. Abbé Moigno.
- Traité Experimental de l'Electricité et du Magnetism, par M. Becquerel.
- Der Electro-magnetische Telegraph, von Dr. H. Schellen.
- Der Praktische Telegraphist, oder die Electro-magnetische Telegraphic nach dem Morse's chen System, von F. Clemens Gerke.

Author's Preface

- Kurze Darstellung der an den Preussischen Telegraphen-Linien mit unterirdischen Leitungen, von Werner Siemens.

Since the following pages have been in type, an Act of Parliament to amend the laws relating to Letters Patent has received the Royal assent, and comes into operation on the 1st of October, 1852. By this enactment many of the evils so bitterly and so justly complained of have been remedied; and although many of the provisions contained therein seem as yet far from perfect, still it is hoped that the Act will prove a great boon, not only to inventors and patentees, but to the community at large.

E. H.

LONDON, JULY, 1852.

1
INTRODUCTION

What an age of wonders is this! When one considers the state of Science a century ago, and compares the light of the past with that of the present day—how great is the change! How marvellous the advance! Discovery has followed discovery in rapid succession—invention has superseded invention—till it would seem to the superficial observer that little now remains to be discovered, and that further improvement is next to an impossibility.

But who shall set a limit to the inventive genius of man, and say, "hitherto shall it go, but no further, and here all discovery and invention shall cease"?

Each discovery brings with it new light. This light illumes many a previously dark and untrodden path, and displays to the astonished gaze a long vista of new wonders, and still greater discoveries. These in their turn enlarge the range of man's mental vision, and further progress follows. But still the darkness around is great indeed. The eye of the mind cannot yet fully penetrate the thick veil which even now conceals the naked majesty of scientific truth, nor is it enabled to comprehend the varied powers of Nature that are still buried in obscurity. As in a dark night a sudden flash of light only dazzles the eye of the observer, so a great discovery at first unnerves the mind, and for a time renders it incapable of further observation. Notwithstanding this, the advance in scientific knowledge has of late been vast and important. That which is now thought chimerical, soon becomes an acknowledged fact, and the marvellous of today is no wonder on the morrow.

Few inventions have created so great surprise and delight as that of the Electric Telegraph.

It is but a few short years since the electric telegraph was classed among the chimeras of the age, and yet already its tiny wires lie stretched across the length and breadth of the land, or buried in the depths of the ocean, and man can now talk with man although a thousand miles apart.

The author having been requested to write a short treatise on this wonderful invention, has consented to do so, in the full belief that the reader will make due allowance for the many imperfections that cannot but exist in a treatise, where, amidst the many duties of his arduous profession, so little time has been allowed for its compilation.

Relying, therefore, on this leniency, the author trusts that what he has written may not be read without interest; and he hopes that the perusal of this little work may induce many to search further among the hidden treasures of Nature, and through Nature's powers to look up with greater reverence to the Being that created them.

2
TELEGRAPHS GENERALLY

TELEGRAPHS BY MEANS OF LIGHT

Long before the electric telegraph had been thought of, the attention of individuals as well as governments had been directed to the obtaining of a rapid communication of intelligence between distant points.

In olden times, fires were used to announce the approach of an enemy, or to telegraph the news of victory or defeat.

The Ancient Britons had their signal fires, to warn the country of the approach of an enemy.

The Romans employed the same expedient in times of war, to telegraph their victories or defeats.

During the wars that raged in Ireland, the news of a victory or defeat was telegraphed across the Irish Channel by fires lighted on Fairhead in Ireland, and on the Mull of Kintyre in Scotland. Help was thus asked for or telegraphed as not needed, as occasion required.

The remains of the stations for these beacon-fires still exist in this and other countries, and the embers seem even now but smouldering in the dust.

It will be in the recollection of many now living, that the approach of the French to this island was to have been announced by the lighting of immense fires on the tops of many of the high hills in England, and the news thus telegraphed from one end of the kingdom to the other in a very short period of time.

Beacon-fires were abandoned when the aerial or semaphore telegraph was invented.

The aerial telegraph was employed for a considerable period, both in England and on the Continent, before the electric telegraph began to assert its claims. Even now its use is continued in some parts until its rival can be erected to supply its place.

TELEGRAPHS BY DIRECT SIGHT

No sooner was the telescope invented in England by Robert Hook, than this instrument was employed for observing signals at a great distance. Men's minds were then naturally turned to the invention of a *set* of signals, so that intelligence of any kind might be read off with great rapidity.

Among the labourers in this field may be mentioned with distinction the names of Amontous, Linguet, Gautley, Hanan, and particularly M. Claude Chappé, a French engineer. M. Chappé's inventions were first tried in the year 1791. Like all inventors, Chappé met with great opposition and discouragement. The people were opposed to the use of telegraphs at all, his first telegraph and the station were destroyed by the populace. His second telegraph shared the same fate, it was burnt to the ground and Chappé himself narrowly escaped with his life. The people threatened to burn him along with his telegraphs.

Subsequently the matter was taken up by the French Government. A commission was appointed and the Commissioners reported favourably on the plans of M. Chappé. His system was ordered to be adopted and Chappé himself was honoured with the appointment of Telegraphic Engineer to the French Government.

The aerial telegraph between Paris and Lille was constructed in 1794 on his plan, and two minutes only were occupied in communicating intelligence between those places. This form of aerial telegraph was afterwards much used on the Continent.

It appears that the Swedes established their first aerial telegraph in 1795; the English in 1795; the Danes in 1802: Asia had one in 1823, between Calcutta and the fortress of Chunore; Mahomet Ali established one between Alexandria and Cairo; Prussia had a similar telegraph in 1832; Austria in 1835 ; and Russia in 1839.

As the electric telegraph advances, this method of communicating intelligence is now falling into disuse. The aerial telegraph between London and Portsmouth has for some time been superseded by the electric telegraph. It is now entirely removed, and

Telegraphs Generally

wires supply its place. The only one now in use in England is from Liverpool to Holyhead, and this will shortly disappear, inasmuch as wires between those points are already laid. The expense of maintaining and working the aerial telegraph is very great. The line from London to Portsmouth cost no less than £3,300 per annum; and the one between Liverpool and Holyhead costs now about £1,500 per annum. In a short time the aerial telegraph will be numbered amongst the things that were. It was with the greatest difficulty, however, that the Lords of the Admiralty could be induced even to try an electric telegraph. In 1823, Mr. Ronalds, of Hammersmith, wrote to the Lords of the Admiralty, requesting an inspection of his electric telegraph. He strongly recommended its adoption for Government purposes. The result of this communication will be best learnt by an extract from a work published by Mr. Ronalds in 1823. Mr. Ronalds in a note says:

> "Lord Melville was obliging enough, in reply to my application to him, to 'request Mr. Hay *to see me on the subject of my discovery*'; but before the nature of it had been known, except to the late Lord Henniker, Dr. Rees, Mr. Brande, and a few friends, I received an intimation from Mr. Barrow, to the effect *'that telegraphs of any kind were then wholly unnecessary, and that no other than the one then in use would be adopted'*.
>
> I felt very little disappointment and not a shadow of resentment on the occasion, because everyone knows that telegraphs have long been great bores at the Admiralty. Should they again become necessary, however, perhaps Electricity and Electricians may be indulged by his Lordship and Mr. Barrow with an opportunity of *proving* what they are capable of in this way."

Such was the assistance rendered by our Government to one of the early inventors in electric telegraphs. It appears that they would neither consent to an interview with the inventor, nor even inspect his telegraph.

The aerial telegraph was capable of being used only for about one-third of the year. Darkness—fogs—and storms of rain or snow, were constantly cutting off all communication. Yet despite all these obstructions, how difficult it was to prevail on the British

Government to make even a trial of the electric telegraph, although this method not only sets at defiance all these interruptions, and conveys its messages with the speed of lightning by night as well as by day,—but carries its intelligence silently, either over the land or under the earth, or far below the troubled waves of the mighty deep.

Who shall say what may yet be done by the agency of electricity? Much has already been achieved, but more, far more, remains yet to be accomplished. Vain were it for man to attempt to foretell the results which may yet be produced by this mysterious power. Many and great are the inventions and discoveries that even now lie concealed in the womb of Time. But the day *will* come when these shall all be brought forth to benefit and bless the community, and man, reviewing with astonishment the progressive results which the workings of his genius and the unwearied exertions of his mind and body have already one by one achieved, will never rest satisfied until all the powers of Nature are made to minister to the wants, the comforts, and the pleasures of himself and his fellow mortals.

TELEGRAPHS BY ELECTRICITY

It is now scarcely a century since electricity was first proposed to be employed for the purpose of communicating intelligence between distant places.

Many and various were the discoveries and inventions that were required before the electric telegraph could be made to assume its present state.

It is not to one person alone that the world is indebted for this wonderful invention. Hundreds of eminent persons have laboured in the field, and numerous men of science have from time to time added their quota both of discovery and invention.

It was not for many years after the *first form* of electric telegraph was devised that sending telegraphic intelligence, by means of electricity, to any considerable distance, could be regarded as either physically or commercially practicable.

Even at this day the electric telegraph may be considered as only in its infancy. Every year improvements are being made, and fresh discoveries come to light.

That considerable proficiency, however, in the art of telegraphing by electricity has already been attained, no one can for a moment deny; still much remains to be done, both scientifically as well as commercially. The construction of the electric telegraph is still to a great extent imperfect, and the charges for its use are at present too high. It is, therefore, no wonder that it has not as yet come into that general and very extended use to which it seems so admirably adapted.

Much use, however, is even now made of the telegraph. The man of extensive business only wonders how he could possibly have done without it so long: still the multitude does not as yet employ it as one of its usual and ordinary means of communication. On extraordinary occasions the telegraph is much resorted to, it is true; but on ordinary occasions, seldom or never.

To give *in extenso* the various discoveries that have been made in this branch of science, and to enumerate the many inventions that have followed, would far surpass the limits allowed on the present occasion.

The three great fundamental principles on which all telegraphs at present in use depend are as follows:

The first is the discovery by Oersted, viz.

1. That a magnetic needle, when in proximity to a body through which a current of electricity is passing, has a tendency to place itself at right angles to such body, or, more strictly speaking, to rotate round that body.

The next is the discovery of the late Mr. Sturgeon, viz.

2. That if currents of electricity pass around a bar of soft iron, the iron becomes temporarily magnetic; and when in that magnetic condition it powerfully attracts to it any pieces of soft iron which may be placed in its vicinity.

The third is based upon the joint discovery of Sir Anthony Carlisle, Mr. Nicholson, and Sir Humphry Davy, viz.

3. That when a current of electricity passes through certain chemical substances, those substances are thereby decomposed, or new compounds are formed.

On these three great fundamental principles all the telegraphs at present in use depend.

The Electric Telegraph itself may be fairly divided into three distinct and separate parts, viz.

1. The means resorted to for producing electricity.
2. The means adopted for rendering a current of electricity cognizable to the senses.
3. The means usually employed for transmitting to a distant place the electricity required.

3

ON THE PRODUCTION OF ELECTRICITY

Electricity may be produced in a variety of ways: by friction; by chemical action; by magnetic induction; by change of temperature; by the power and at the will of certain animals. The three first modes are those usually employed in the electric telegraph.

Electricity from Friction

The discovery of frictional electricity is of very ancient date. Thales of Miletus, who lived about 600 years before the Christian era, is reported by subsequent writers to have described the power developed in amber by friction, by which the amber was enabled to attract to it pieces of straw and other light substances.

Theophrastus (321 B.C.), in his writings, describes this property. Pliny (A.D. 70) refers also to the same. He speaks of pieces of amber that *"attritu digitorum acceptá vi caloris attrahunt in se paleas et folia arida ut magnes lapis ferrum."* †

Similar remarks may also be found in the writings of Priscian and Solinus.

Salmasius, in his commentary upon Solinus, asserts that the word *karabe*, by which amber was known among the Arabs, is said by Avicenna to be of Persian origin, and to signify the power of attracting straws.

It does not appear that any of the ancients reasoned upon those observed effects; they merely observed, and recorded them as facts.

Dr. Gilbert, however, at the commencement of the sixteenth century, instituted a series of experiments upon the subject. He

---

† **Editor's Note:** Latin translation: ...that *"by friction and heat of the fingers, attract both straw and the leaves of the dry land, like the lodestone attracts iron."*

found that the property possessed by amber was not confined to that substance alone, but belonged to several other bodies, such, for instance, as the diamond and many other precious stones, glass, sulphur, sealing wax, resin, &c.

Boyle found that warming these bodies increased the effect.

To Otto Guericke, of Magdeburg, who was also the inventor of the air-pump, is due the invention of what is commonly called the Electrical Machine. This philosopher mounted a globe of sulphur upon an axis, and, on turning the globe round, applied friction to it. By this means he detected that strong electrical excitation was accompanied both by light and sound: he also discovered that after a body had been electrically excited, and another light body brought in contact with it, a repulsion ensued. Many other of the now well known phenomena of attraction and repulsion were demonstrated and recorded by this philosopher.

In 1675, Sir Isaac Newton made several important discoveries relating to the above, and noted down the effects observed in his experiments on the subject.

Hawkesbee, between 1705 and 1711, made various discoveries. He substituted a globe of glass in lieu of the globe of sulphur of Otto Guericke, and fixing it in a wooden frame, he produced an electrical machine very similar to those now in general use.

Grey and Wheeler experimented further, and succeeded in producing motion in light bodies at distances of 666 feet.

M. Du Fay, between 1733 and 1737, conducted a series of important experiments, and greatly enlarged the number of phenomena observable in bodies when acted on by electricity in this state.

The Abbé Nollet, on witnessing some experiments, discovered a simple principle that accounted for the apparently anomalous results obtained by former experimenters, and explained satisfactorily the cause of a body being first attracted and then immediately repelled after contact.

Amongst those who about this time laboured in the science may be mentioned with distinction the names of Desaguliers of France, Boze of Wittemburg, Winkler of Leipzig, Ludolf of Berlin, and Dr. Miles; each of whom brought fresh facts to light, or improved upon the apparatus then in use.

Dr. Watson, in 1745 *et seq.*, conducted several important experiments, which are duly recorded in the "Philosophical Transactions."

Kleist and Muschenbroeck, at Leyden, simultaneously discovered a means of accumulating the electric power by the invention and employment of the Leyden jar, although the honour of this discovery is by some attributed to a person named Conesus.

Dr. Bevis recommended the coating of the outside of the jar with tinfoil, water having been previously used by Muschenbroeck and Kleist in the interior of the jar.

Dr. Watson, however, applied the tinfoil both to the inside as well as the outside of the jar, and thus perfected the Leyden jar. In this state it now remains.

The distance to which the electric power might be conveyed next occupied the attention of philosophers both in England and France. Experiments were made in the Tuileries on the subject, and electricity transmitted through a circuit of considerable length.

Dr. Watson, in 1747, in the presence of many scientific persons, transmitted the power through 2,800 feet of wire and 8,000 feet of water, thus employing in his experiments the use of the *earth circuit*.

Afterwards, on the 14th of August, 1747, Dr. Watson conducted an experiment on a much larger scale at Shooter's Hill. The wire was insulated by baked wood, and was 10,600 feet or nearly 2 miles long.

But as we are now trespassing on that part of the subject which consists of the means employed for conducting the electric power to a distant point, we must return to the production or generation of the electric power itself.

From this period until the invention of the Hydro-Electric Machine little progress was made in the art of producing electricity by friction.

In 1840, at Newcastle on Tyne, it was observed that a jet of steam issuing from the boiler of a steam engine emitted electricity in considerable quantities, and that on applying a conductor to the jet, powerful sparks were obtained.

Mr. Armstrong, of Newcastle, made many experiments on

this newly discovered source of electricity, and ultimately constructed what is now called the Hydro-Electric Machine.

The researches of Dr. Faraday brought to light the fact that the electricity developed by this machine was due to the friction of the watery particles in the steam on the sides of the surface. An enormous amount of power may be developed by means of this apparatus. Not only have gun powder and shavings of wood been set on fire by the spark direct from the machine, but in some experiments it has also been found, that when a current was sent through a galvanometer the needle was deflected between twenty and thirty degrees, and iron converted into an electro-magnet.

This brings the inquiry down to the present time, with reference to the production of electricity by friction.

We will now proceed to describe the discoveries that have been made in the production of electricity from chemical action.

ELECTRICITY FROM CHEMICAL ACTION

After the Leyden jar and the electric battery, composed of a number of such jars, had been experimented on in various ways, and by means of the power so accumulated metals had been fused, volatilized, reduced to dust, or dispersed in air, the lives of animals and vegetables taken away, and other striking effects produced on matter, for a long period little or no further progress was made.

At length Galvani, in 1791, stumbled as it were by chance upon what was then thought a new fact in the science. This ultimately led to most important consequences. Through it, means were obtained of producing enormous quantities of electricity, and that from the chemical action of bodies on each other.

It appears that Du Verney had made the very same observation as Galvani had done about a century before, without the circumstance having attracted the attention of philosophers at the time.

The reader will probably be too well acquainted with the story of Galvani and the frogs to need a repetition of the circumstance. Suffice it to say, that the accidental contraction of the muscles of a frog, when in the proximity of an electrical machine, led to some of the most brilliant discoveries that have ever adorned the annals of science.

On the Production of Electricity

Various hypotheses were framed to account for the peculiarities observed in the experiments with the muscles of animals.

Valli wrote on the subject, and in 1793 Dr. Fowler published his essay on *Animal Electricity*. The same subject was also investigated by Dr. Robison.

Professor Volta, of Pavia, confuted many of the theories adduced, and ultimately produced the arrangement known as the Voltaic Pile, the first rude form of what is now termed the Galvanic Battery. A letter of Volta on the subject was published in the "Philosophical Transactions" for 1793.

During the heat of the discussions between the partisans of the theories of Galvani and Volta, Fabroni repeated many of the experiments, and communicated his researches to the Florentine Academy.

It is in this paper that the first suggestion as to the *chemical origin* of galvanic electricity is to be found.

On the 20th of March, 1800, Volta addressed a letter to Sir Joseph Banks, then President of the Royal Society, in which he announced to him the discovery of the voltaic pile.

After due investigation of this instrument, Volta endeavoured to improve the arrangement of its parts, in order to obtain a greater amount of power. The result was the invention of the apparatus known by the name of *La Couronne de Tasses*. This arrangement consisted of a circle of cups, each cup being filled with warm water or with a solution of sea-salt, and having also a piece of silver and a piece of zinc in the liquid.

The pieces of the two different metals in the same cup were not in metallic contact, but the zinc of the one cup was metallically united to the silver of the *adjacent one*, and so on throughout the series, the liquid alone intervening between the metals in the *same* cup. Thus it is evident that in this arrangement of the *Couronne de Tasses* we have a complete and perfect galvanic battery. We have the insulated cell, and the two metals in the cell separated by a liquid capable of acting chemically upon one of them.

Many important improvements have, however, been made in the materials employed, though the principle of the battery remains now as it left the hands of Volta.

Dr. Wells, in 1795, discovered that charcoal might be substituted for one of the metals in the cells.

Mr. Cruikshank, of Woolwich, in 1800, arranged another form of battery, making the metals employed to form divisions of the trough.

In Dr. Wollaston's arrangement, in 1815, dilute sulphuric acid was employed, and two pieces of copper were used to one piece of zinc, the copper being placed on each *side* of the zinc plate. He arranged the whole in a trough composed of a number of cells, and attached the plates to a rod so that the whole of the plates might be lifted out of the liquid when the electric power was not required. By this arrangement less waste of the zinc and acid resulted, as the zinc was not being dissolved by the acid when the battery was out of action. It will be observed, however, that the above are but mere modified forms of Volta's *Couronne de Tasses*.

Many experiments were made by Valli, Fowler, Robison, Dr. Wells, Humboldt, Fabroni, Nicholson, Carlisle, Cruickshank, Haldane, Henry, Davy, Wollaston, Trommsdorff, Van Marum, Pfaff, Aldini, Hisinger, Berzelius, De Luc, De la Rive, Becquerel, and others; and new facts were added yearly to the existing stock of knowledge. Several kinds of acids were used; charcoal was substituted for one or both of the metallic plates; wires were made red-hot, and various substances difficult of decomposition easily decomposed by the electric action.

Other powerful forms of galvanic batteries were afterwards contrived, both by Professor Daniell and Professor Grove—forms and arrangements which admit of a uniform and continuous flow of the electric power for a considerable period of time.

Daniell's constant battery, as it is called, consists in having two liquids in each cell, the liquids being separated by a porous diaphragm—the one liquid being dilute sulphuric acid, and the other a saturated solution of sulphate of copper: in the latter, copper plates are immersed, and in the former, plates of zinc.

Grove's battery consists of two liquids and two metals—the liquids being nitric acid and dilute sulphuric acid, and the metals platinum and amalgamated zinc: the plates of platinum are immersed in the nitric acid, and the zinc in the dilute sulphuric acid.

An important discovery, in order to prevent the local action of the diluted sulphuric acid on zinc, was made by Mr. Sturgeon and Mr. Kemp. This consisted in rubbing mercury over the surface of the zinc. By this means the other forms of galvanic batteries are made to last a much longer time, and the flow of electricity during the action of the battery becomes far more constant and regular.

The relation which the galvanic battery bears to the Leyden jar or the common electrical machine may be thus stated:

> In the Leyden jar, a sudden, instantaneous, and violent effect is produced on any body through which the power passes: a torrent of force precipitates itself, as it were, instantly along the line of communication, while in the galvanic battery the power flows in a gentle and continuous stream, producing a constant and uniform action for any definite period of time.

From the Leyden jar, the whole force passes in an inconceivably short space of time, while from the galvanic battery the action may be continued as long as desired.

Just as in mechanics a sudden blow from a hammer differs from a continued pressure, so does the action of electricity from the Leyden jar differ from that produced from the galvanic cell.

During the last few years improvements have been made in batteries, and especially with regard to the peculiar requirements of the electric telegraph.

Mr. Cooke uses sand mixed with the diluted acid, in order to prevent, to a certain degree, the mechanical transmission of the salts of the one metal on to the surface of the other.

Messrs. Brett and Little employ a gradual renewal of the diluted sulphuric acid, and allow the exhausted liquid to run out from the bottom of the cells and fresh liquid to drop into the cells from the reservoir above.

The author, however, employs, instead of diluted sulphuric acid as the exciting liquid, a solution of the sulphates of the earths, such as the sulphate of magnesia, or the sulphate of alumina. By this arrangement, not the slightest attention need be paid to the batteries required for working the telegraph, for many months together: a constant and equable flow of the power may be thus obtained whenever desired, and that for a whole year, although, in some instances, 200 messages a day have been transmitted. Another advantage is, that one liquid only need be used, and that of the cheapest kind.

ON THE PRODUCTION OF ELECTRICITY FROM THE MAGNET

The honour of this discovery is due to the illustrious Faraday,—Dr. Faraday having noticed that on bringing one wire through which a current of electricity was passing near to another wire, a current of electricity was induced in the second wire; and that when the connection between the poles of a galvanic battery was broken, a current passed for an instant through the circuit in a direction *contrary* to that in which the current was proceeding before.

Further research into these curious phenomena ultimately led this philosopher to the production of *electricity from magnetism.*

The first apparatus employed was a ring of soft iron, around one half of which an insulated wire was wound connected with a battery, and around the other half of the ring another insulated wire was wound, but not in metallic connection with the former one: on passing a current through the first wire the ring of iron became magnetic, and at the same instant that the iron was assuming this magnetic state a current of electricity was found to be traversing the *second* wire, although it was not in contact with the first-named wire. This secondary current, however, ceased almost immediately, but was again renewed in the opposite direction when the ring was losing its magnetic state, by disconnecting the battery from the first-named wire. This inverse or induced current lasted only until the iron re-assumed its normal condition.

Thus it was evident to Dr. Faraday, that if a piece of iron be surrounded with insulated coils of wire, a current of electricity might be induced in such wire by merely magnetizing and demagnetizing the iron.

An apparatus was therefore constructed by means of which the iron so wound with wire was magnetized and demagnetized, by being made to approach to and recede from a permanent magnet. The result was as anticipated, and powerful currents of electricity were thereby induced and traversed the circuit appointed for them.

The iron was now removed from the helix of wire, and a permanently magnetic core *introduced* and *withdrawn*,—at each introduction and withdrawal of the magnet a current of electricity was developed in the wire.

A copper disc was then employed and made to rotate in the presence of a magnet,—currents of electricity were at once detected in the disc.

By these means water was decomposed—magnetic needles moved—iron and steel magnetized—and all the other well known effects of electricity, both frictional and galvanic, were produced.

Amongst the labourers in this branch of science may be reckoned the names of Babbage, Herschel, Barlow, Nobili, Antinori, Bacelli, Christie, Prevost, Colladon, Harris, and others.

Various forms of magneto-electric machines were subsequently devised.

Dr. Faraday, however, was the first person who constructed one by means of which a continuous current of electricity was produced.

In 1832, M. Pixii, of Paris, constructed a machine in which a coil of copper wire was employed instead of a disc of copper; and in 1833, Mr. Saxton submitted to the British Association his magneto-electric machine, in which the coils rotated and the magnet was fixed.

Another form of the same kind of machine was executed by Mr. Clarke, and is fully described by M. Becquerel.

From this period up to the present time little further improvement has been made in producing electricity from the action of the magnet.

In 1837, Dr. Steinheil used the ordinary magneto-machine for working his printing and other telegraphs.

In 1841, Professor Wheatstone patented a mechanical arrangement by means of which, with a series of electro magnets rotating before a magnet or a set of magnets, a continuous current of electricity in the *same* direction was produced. This instrument was intended to supersede the use of galvanic batteries in electric telegraphs, but it does not appear to have ever been brought into practical operation.

A modification of Mr. Saxton's arrangement is, however, now employed to a considerable extent in producing electricity for the ringing of the alarums of a telegraph. It should be observed with regard to this method of producing currents of electricity, that electricity is generated in the conducting wire only when an adjacent electro-magnet is undergoing a *change* in its magnetic state; and that when the magnetism of such electro-magnet is *increasing*, the current of electricity developed is in one direction, whilst, when the magnetism is *decreasing*, the current of electricity is in the *opposite* direction; and also, that when the magnetism of the electro-magnet is *stationary*, *no* current of electricity ensues.

This method of producing electricity through the medium of the magnet is well adapted for causing the liberation of the detent of wound-up mechanism, where the power used is required only for a very short period of time.

The next subject for inquiry is:

PRODUCTION OF ELECTRICITY BY A CHANGE OF TEMPERATURE

Professor Seebeck, of Berlin, was the first to observe that a current of electricity could be generated by joining together arcs of two different metals, and applying heat at either of the points of junction. This discovery was no sooner known than many philosophers repeated and extended the experiments made by Professor Seebeck.

Amongst the first labourers in this field were Baron Van Zuylen, Dr. Van Beck, and Professor G. Moll.

Many were the experiments made by these philosophers, and great additional knowledge was obtained by them in this new branch of the science of electricity.

Of the details of the experiments of Professor Seebeck little was known in England at the time.

Professor Cumming was the first to undertake the task of investigating thoroughly the peculiarities of this new species of electricity, and to him science is indebted for several very important facts developed in his experiments.

Professor Cumming clearly proved, that all that was necessary in order to produce in a conductor a current of electricity, was, that one end of the conductor should be at a different temperature from that of the other end.

It matters not whether the temperature of the one end be raised by the application of heat, or whether it be lowered to an equal degree by the application of cold. A current of electricity in both cases is produced, the only difference being in the *direction* of the current,—the current at all times progressing in one and the same direction with regard to the relative position of *the hotter and cooler ends* of the conductor. Various forms of thermo-electric apparatus have been devised, and the power multiplied by the repetition of alternations of the pieces of metal employed,—in manner somewhat resembling the repetitions of the alternations of the elements in the galvanic battery.

The two metals which, when their ends are soldered together, produce the most powerful current of electricity, are antimony and bismuth.

A detailed account of the ingenious experiments of Professor Cumming will be found in the "Transactions of the Cambridge Philosophical Society" for 1823.

The author does not deem it necessary here to dwell longer upon this mode of producing electricity; it has not been as yet practically applied to the electric telegraph.

The field is one rich in the extreme, and from which, no doubt, in time, will be extracted great knowledge as to the constitution of matter, and the action of particle on particle; nor is it at all improbable that the day is not very distant when a farthing rushlight will be made capable of developing sufficient electricity to keep up an instantaneous communication by telegraph between London and Liverpool.

Electricity from Animals

Another source whence currents of electricity may be obtained is to be found in certain animals.

Although this is a subject which does not properly belong to a treatise on the electric telegraph, yet it has been thought well briefly to allude to it, simply for the purpose of enumerating the various sources whence electricity may be derived, and to show how admirable the contrivance of Nature is, when she desires to produce electricity in the midst of masses of water, and to confine the current to a particular course.

Various instances are recorded in ancient works of sparks of electricity having been given off from the hair and other parts of the human body.

Virgil makes mention of a harmless flame which was emitted from the hair of Ascanius.

Whether this was so or not is uncertain, as doubts even on the existence of such a person as Ascanius have been raised.

Several ancient authors relate the same thing, however, of Servius Tullius, the Roman king. Numerous similar instances might be adduced, but it would be out of place here to enumerate them.

Another fertile source of electricity, recorded by various writers, is to be found in the human body on taking off a part of the dress. The electricity so produced has been the subject of much investigation, and has been tested by the electroscope in a variety of ways. In all these latter cases, however, the cause seems to have been rather in the friction of the garment upon the skin or hair. These effects therefore should hardly be enumerated under the head of Animal Electricity.

There is, however, undoubtedly a very fertile source of electricity in the bodies of certain fishes, produced independently of friction. This power is often very intense, so much so as to deprive smaller animals of their life, and to stun or render powerless for the time being even larger animals, such as the horse.

Amongst the most remarkable of this class of animals may be enumerated the raia torpedo, gymnotus electricus, silurus electricus, trichiurus electricus, and tetraodon electricus.

Redi, in 1678, gives an account of certain imperfect observations of fishermen on the raia torpedo.

Kempfer, in 1702, also describes some experiments he made on the same fish.

Mr. Walsh, in 1773 and 1775, published some important papers on this subject, and was the first to prove the electrical origin of the power emitted. In one of his experiments he caused the shocks to be transmitted through a chain of five persons.

In 1773, Dr. Ingenhous made other similar experiments.

Spallanzani also followed in the same course of experiments, and proved that even the foetus of the torpedo possessed this remarkable property.

More recent experiments were made by MM. Humboldt and Gay-Lussac. They have proved that the least injury to the brain of the animal destroyed at once its power of producing these shocks. A series of numerous facts as to the passage and peculiarity of the power are recorded by them.

In 1773, the celebrated Hunter published, in the "Philosophical Transactions," the anatomical structure of this fish, showing the position of the electric organs.

In a fish 18 inches long, it was found that the number of columns composing each organ amounted to 470.

In a very large torpedo, found on the British coast, which was 4½ feet long, and weighed 73 pounds, the number of columns in each organ amounted to 1,182—a battery power of no despicable order.

It appears also, that the magnitude and number of nerves in this animal are very far greater than those supplied to any other animal whatever; and it is also evident from the experiments, that the power of transmitting these powerful shocks of electricity is controlled and regulated by the will of the animal.

Similar remarks may be applied to the other fishes above mentioned.

Among the list of labourers in these investigations may be mentioned the names of M. Richer, Redi, Schilling, Dr. Hunter, Mr. Walsh, Dr. Williamson, Humboldt, and M. Geoffroy.

The subject of the spontaneous production of electricity by animals would not have been introduced in this work except for the purpose of showing the very remarkable fact, that, however difficult it is found in practice for man to transmit, artificially, currents of electricity from any kind of electric apparatus wholly submersed in water, yet Nature, in her sublime workings, finds no difficulty whatever in so doing. The philosopher is thus invited to careful study and deep investigation. The day may come when this mode of action in the animal kingdom will be better understood than it is now, and then probably will be discovered a means of constructing submarine telegraphs, without any insulation of the wires: and who shall say whether such a discovery would not satisfactorily solve the problem of communicating instantaneously between Great Britain and America?

Electricity from Other Sources

The foregoing are the principal sources whence electricity in large quantities may be obtained, but they are by no means the only ones. This subtle power of Nature may be produced by a variety of other methods.

MM. Lavoisier and La Place have demonstrated that bodies in passing from the solid or the liquid state to that of vapour, and conversely from the vaporous to the liquid or solid state, show unequivocal signs of the presence of electricity.

Another source of electricity is to be found in the pressure of certain crystals.

Thus, for instance, Iceland spar, on being subjected to pressure in certain directions, shows evident symptoms of electric action, and affects the galvanoscope accordingly.

Amongst the various bodies which possess this property are Iceland spar,—topaz,—uclase,—arragonite,—fluate of lime,—carbonate of lead,—elastic bitumen, and other substances.

The atmosphere is also another very fertile source whence electricity may be derived.

Setting aside those magnificent natural displays of the electric action, viz. the thunder-storm and the aurora borealis, not a cloud passes over our heads but the electric equilibrium of the earth below is affected thereby.

In some states of the weather, and in certain fogs, an insulated rod extending high into the atmosphere, with a range of exploring wire attached to it, will bring down torrents of the electric fluid. It is a remarkable fact too, that the power within a few minutes changes from positive to negative, and, *vice versa*, from negative to positive.

Whatever may be the nature of this marvellous power, we know not. We observe its effects,—we record its actions,— we test its presence,—we find it everywhere, but we know not what it is. Scarcely an atom of matter can move from its fellow atom but an electric current may be detected; and yet, of the nature of this current we are entirely ignorant.

The above is a brief description of a few of the many sources of obtaining that power which circulates as the life blood in the electric telegraph. To have attempted to have gone further into this part of the subject would have been out of place, and would only have extended this little treatise beyond the limits allowed, without in any way rendering it possible to have done justice to the investigation.

Those who wish to know more, either theoretically or practically, in this most interesting science, are referred to the many excellent works already published on the subject.

It is proposed, therefore, to pass on to the consideration of the means employed for making the presence of electricity discernible at a distant point.

4
METHODS FOR DETECTING ELECTRICAL SIGNALS

It would perhaps appear to have been more regular if the means employed to convey the current to a distant point had been now considered, instead of proceeding to discuss the expedients employed at that distant point to record the presence of the current.

But it was not until the effects produced by electricity were well known, that the desire appeared to ascertain how great a distance might intervene between the generation of the power and the effects produced.

It is thought therefore best to follow this course in the following remarks on the electric telegraph: the more so is this deviation justified, as this order was in fact the one in which the invention of the electric telegraph was made.

No sooner had the sulphur ball of Guericke, the glass globe of Hawkesbee, and the glass cylinder of Gordon, been used, than ready means were furnished of trying experiments on an extended scale; and when the Leyden jar was invented, the apparatus for research became almost perfect.

The very first notice that we have of the effects produced by electricity consists in its power of attracting light substances, as bits of straw, &c. Such effects are said to have been described by Thales, some 600 years before the Christian era.

To Dr. Gilbert, however (A.D. 1600), is due great credit for the multiplicity of his experiments on this head.

In order to test the effects of the various bodies, Dr. Gilbert brought them to the end of a light needle of any metal, balanced and turning freely on a pivot, like a magnetic needle.

Otto Guericke clearly demonstrated that a light body, after it had by attraction been brought into contact with an excited electric, was repelled by it. He also found that if light bodies were suspended within the sphere of action of an excited electric, they themselves became possessed of electrical excitation.

Now this property is one which was employed in one of the first electric telegraphs at work in this kingdom, viz. in Ronalds': in this telegraph two pith-balls were made to diverge by electricity when desired, and thus to denote the signals desired.

It should be remarked, that up to the year 1720, attraction and repulsion were considered as the only absolute proofs of the presence of electricity, although it had also long been observed that light was produced by electrical excitation.

For some time the minds of philosophers seemed to have been devoted to the production of as powerful an electric spark as possible. Thus we have Boze, Winkler, Gordon, and Ludolf of Berlin, all labouring with this end in view,—the principal object appearing to be to set inflammable substances on fire thereby.

Ludolf of Berlin was the first to accomplish this: he succeeded in setting on fire a highly inflammable spirit.

The noise made by this firing of a spirit was employed in 1816 in Ronalds' electric telegraph, for the purpose of calling attention previously to the communication of intelligence.

The passage of the electric spark was used by Reizen, in 1794, as the means of designating any of the letters of the alphabet.

Another mode, in which frictional electricity was proposed to be used, was, by separating the conducting wire and passing a strip of paper uniformly between the severed ends thereof. When a charge from a Leyden phial or battery was passed through the circuit, a hole was pierced through the paper.

The Rev. H. Highton, in 1844, proposed and patented a plan of telegraph on this principle.

We will now pass on to the means used when galvanic electricity was employed for rendering its presence discernible by the senses. (Oersted, in 1819, discovered that a magnetic needle delicately suspended in proximity to a conductor through which an electric current was passing had a tendency to place itself at right angles to such a conductor. The application of this principle to the electric telegraph has been almost universally adopted in this kingdom, and most extensively employed in other parts of the world.

To Ampére is due the discovery that a wire through which a galvanic current is passing may be made to assume all the properties of the magnet itself. This induced magnetic power was found to cease the instant that the current was arrested.

M. Arago also, at the same time, published the fact that iron filings were attracted by such galvanic wire, and that the wire had thereby the power given to it of producing temporarily, in iron, magnetic properties that did not previously exist in it. In this way M. Arago showed that the wire had the effect of permanently magnetizing a needle of steel.

Many of these experiments were conducted jointly by MM. Arago and Ampére. These philosophers investigated the action of coils and helices of wire, and at length demonstrated that a helix of wire with a current of electricity passing through it may be made to produce all the effects of the magnet itself.

Sir Humphry Davy at this time commenced a long series of experiments, which have proved of the greatest value to the science. Very few new facts, however, seem to have been brought to light by him which have special reference to the effects produced by the electric currents in order to mark its presence on its passing through a conductor, and to make the same cognizable to the senses.

In 1821, Faraday commenced a series of most important experiments. With the advantages he possessed in having at his command a most extended and powerful apparatus, he produced results highly beneficial to the advancement of the science.

The great discoveries of Faraday, at this date, were confined principally to the relative directions and powers of the electric and the induced magnetic forces.

About this period many German philosophers and others repeated and extended the experiments of Oersted and Ampére, —amongst whom may be enumerated the names of M. le Chev, Yelin, M. Brockinan, M. Van Beek, M. De la Rive, M. Moll, Mr. Barlow, Mr. Cumming, and others.

Mr. Cumming, in April, 1821, appears to have been the first

to notice the *increased effects* of a convolution of wire around the magnetic needle, and thus to produce the arrangement known as the galvanometer. This arrangement was subsequently adopted by Professor Wheatstone in his first electric telegraph. The discovery of one of the most important parts, however, of the electric telegraph remains yet to be described.

To the late Mr. Sturgeon is due the discovery of the *electro magnet.* Mr. Sturgeon was the first to discover that if a bar of *soft iron* be surrounded with coils of wire, and an electric current be transmitted in the same direction through each convolution, that the soft iron bar instantly becomes a magnet, and is capable of attracting other pieces of soft iron or steel, and that it remains magnetic so long as the electric current is passing through the coils; and that as soon as the current ceases, the bar instantly loses its magnetic condition, and no longer attracts pieces of adjacent iron or steel.

This property of iron becoming magnetic under the above conditions has entered more or less into almost every form of electric telegraph since the above period, and is one of its most valuable component parts.

Another, and perhaps not less important, effect produced by the electric current, as applicable to telegraphic purposes, is the *decomposition* of water and other similarly constituted substances, when a current of electricity is made to pass through them.

Dr. Lardner, in his treatise on Electricity, says:—

"The invention of the pile had been scarcely more than hinted at, when that course of electro-chemical investigations began, which soon led to the magnificent discoveries of Davy, and the series of experimental researches which have been continued up to the present time with results so remarkable by those who succeeded to him. The first four pages only of the letter of Volta to Sir Joseph Banks were despatched on the 20th March, 1800; and as these were not produced in public till the receipt of the remainder, the letter was not read at the Royal Society, or published, until the 26th June following. The first portion of the letter, in which was described generally the formation of the pile, was shown in the latter end of April by Sir Joseph Banks to some scientific men, and amongst others, to Sir Anthony (then Mr.) Carlisle, who was engaged at the time in certain physiological inquiries. Mr. W. Nicholson (the conductor of the scientific journal known as *Nicholson's Journal*) and Carlisle constructed a pile of seventeen silver half-crown pieces alternated with equal discs of copper and cloth soaked in a weak solution of common salt, with which, on the 30th of April, they commenced their experiments. It happened that a drop of water was used to make good a contact of the conducting wire with a plate to which the electricity was to be transmitted; Carlisle observed a disengagement of gas in this water, and Nicholson recognized the odour of hydrogen proceeding from it. In order to observe this effect with more advantage, a small glass tube, open at both ends, was stopped at one end by a cork, and being then filled with water, was similarly stopped at the other end. Through both corks pieces of brass wire were inserted, the points of which were adjusted at a distance of an inch and three-quarters asunder in the water. When these wires were put in communication with the opposite ends of the pile, bubbles of gas were evolved from the point of the negative wire, and the end of the positive wire became tarnished. The gas evolved appeared to be hydrogen, and the tarnish was found to proceed from the oxidation of the positive wire. It was inferred that the process in which these effects were produced was the decomposition of the water. This took place on the 2nd of May, shortly after the receipt of the first portion of Volta's letter."

Experiments were then tried with wires of various metals, and when platinum wire was used, oxygen gas was evolved from the one wire, and hydrogen from the other. Thus was the decomposing power of the pile established in a few weeks after the discovery of the pile itself. These experiments led Cruickshanks, of Woolwich, to repeat and extend them. On his using litmus paper, the colour of the paper was changed by the current.

This peculiarity possessed by the electric current, of changing by decomposition the colour of bodies submitted to its action, has been variously employed in the electric telegraph in recording the transmission of electric currents, and has now become one of the means of carrying on a correspondence by means of electricity.

Sir Humphry Davy also greatly extended by his valuable researches our knowledge of and insight into this remarkable peculiarity of the action of electricity.

Having now briefly described the progress of discovery with respect to various methods of enabling parties *to record temporarily or permanently the presence of an electric current,* we will now pass on to a few brief remarks on the means employed for conducting currents of electricity from one point to another point situated at a distance therefrom.

ON THE MEANS USUALLY EMPLOYED FOR TRANSMITTING ELECTRICITY TO A DISTANT PLACE

It is evident how comparatively valueless for the purpose of telegraphing would have been the many wonderful discoveries in the production of electricity, if means were not to be found for conveying the power with little or no impediment or loss to a point remote from its source or origin.

This part of the subject for a long period occupied the attention of many of those engaged in the science.

It was found, after many careful experiments, that several substances had the property of conveying electricity through them with but a very slight impediment to its passage.

The metals were found to rank highest in this property. It has been subsequently discovered that all bodies are *conductors* of electricity, more or less. No substance is at present known which is an absolutely perfect non-conductor. With all bodies, the passage through them of a *definite amount* of electricity is but a question of *time.*

The great object to be obtained in the construction of an electric telegraph is, to give the greatest possible facility for the passage of the power to a particular distant station, and to throw every possible obstacle in the way of the escape of any portion of the power in any other direction than the one desired.

For such purpose, the most perfect conductors are used for the conveyance of the power, and the most perfect insulators made to surround such conductors.

The following Table exhibits the conducting power of several

Methods for Detecting Electrical Signals

bodies with respect to electricity. It begins with the most perfect conductors and ends with those which are the least perfect conductors. The properties, therefore, of these latter bodies approximate most closely to that of non-conductors or insulators. The exact order, however, is by no means fully substantiated as yet, and the Table must therefore only be taken as a general guide.

All the metals, viz.

Silver,	Zinc,
Copper,	Tin,
Gold,	Platinum,
Brass,	Palladium,
Iron and	Lime,
Lead,	Dry chalk,
Well-burnt charcoal,	Native carbonate of barytes,
Plumbago,	Lycopodium,
Concentrated acids,	Gum elastic,
Powdered charcoal,	Camphor,
Dilute acids,	Some silicious and argillaceous
Saline solutions,	stones,
Metallic ores,	Dry marble,
Animal fluids,	Porcelain,
Sea-water,	Dry vegetable bodies,
Spring-water,	Baked wood,
Rain-water,	Dry gases and air,
Ice above 13° Fahrenheit,	Leather,
Snow,	Parchment,
Living vegetables,	Dry paper,
Living animals,	Feathers,
Flame,	Hair,
Smoke,	Wool,
Steam,	Dyed silk,
Salts soluble in water,	Bleached silk,
Rarefied air,	Raw silk,
Vapour of alcohol,	Transparent gems,
Vapour of ether,	Diamond,
Moist earths and stones,	Mica,
Powdered glass,	All vitrifications,
Flowers of sulphur,	Glass,

Dry metallic oxides,	Jet,
Oils—the heaviest the best,	Wax,
Ashes of vegetable bodies,	Sulphur,
Ashes of animal bodies,	Resins,
Many transparent crystals, dry,	Amber,
Ice below 13° Fahrenheit,	Shellac.
Phosphorus,	

Since the above Table was arranged, gutta percha has been discovered, and has been found to be one of the most perfect of the so-called non-conductors.

Dr. Desaguiliers devoted considerable attention to this part of the subject, from the time of the labours of Grey until the year 1742. He was the first who applied the term *conductors* to bodies through which electricity passed with comparative freedom. He showed also that the conducting power of animal substances was due to the fluids that they contained.

Dr. Watson also proved, experimentally, that a shock could be passed with great facility through a great number of men at the same instant of time.

The attention of philosophers was now directed to ascertain to *what distance* the shock could be transmitted.

At Paris, M. Nollet transmitted a shock through 180 soldiers. He also formed a chain measuring 5,400 feet by means of iron wires extending between every two persons: the whole company received the shock at the same time.

A discharge from the Leyden jar was also effected through circuits of 900 and 2,000 toises† in length, and in one experiment the basin of water in the Tuileries formed part of the circuit. It was in England, however, that experiments on this subject were made on a more extended scale.

Dr. Watson stretched a wire across the Thames over Westminster Bridge. One of the extremities of this wire communicated with the exterior of a Leyden jar, and the other was held by a person in one of his hands, while the other hand grasped an iron rod. Another person on the opposite side of the river grasped a wire com-

---

† **Editor's Note:** A *toise* is a unit of measure (now archaic), for length, area and volume originating in pre-revolutionary France. Roughly equal to 2 metres.

municating with the interior of the jar. The moment the first-named person dipped the rod into the river, the current passed, and both persons received the shock. This appears to be the first time that a circuit composed partly of wire and partly of the earth was used for transmitting currents of electricity.

The next experiment was made by Dr. Watson at Stoke Newington, near London, where a circuit of nearly two miles was used. This circuit, as in the former case, was made up partly of wire and partly of the earth, the wire being in one case 2,800 feet long, and an equal distance intervening through the earth. It was found, too, that the effect was the same whether the rod was only dipped into water or driven into the earth.

Similar experiments were tried at Highbury in 1744, and finally at Shooters' Hill in August, 1747. In the experiments at Shooters' Hill the wire was 10,500 feet long, the observers being thus separated by a distance of two miles. The wires were supported on *posts* of wood. The whole circuit was therefore four miles long, being composed of two miles of wire and two miles of earth.

It now became a well-known fact that electricity *could* be transmitted over a very considerable distance by means of an insulated wire, and that the effects produced in every part of the circuit were, if not absolutely instantaneous, yet practically so to all intents and purposes. No more experiments were therefore needed to confirm these simple facts. It was absolutely necessary, however, for these facts to be proved, before an electric telegraph could be treated as practically possible.

In 1837, Dr. Steinheil used no less than 7 miles of wire for his telegraph at Munich.

In 1839, Dr. O'Shaughnessy conducted an extensive series of experiments in India, with the view to ascertain the most suitable form of electric telegraph for that country. To Drs. Steinheil and O'Shaughnessy is due the carrying out of Dr. Watson's method, now so generally adopted in Great Britain and America, viz. of suspending the telegraphic wires in the air from post to post. Dr. O'Shaughnessy erected for his telegraphs no less than twenty-two miles of wire, the wires were of iron. They were fastened to poles of bamboo, fifteen feet out of the ground, and were made to hang at distances from each other of about twelve inches. Dr. Steinheil had also 7 miles of wire, which was partly of copper and partly of iron. In Dr. Steinheil's telegraph the wires were four feet one inch apart.

These important experiments of Dr. Watson, Dr. O'Shaughnessy, and Dr. Steinheil set the matter completely at rest, and rendered the idea of communicating intelligence between distant points, by means of electricity, no longer chimerical or doubtful, but a matter of absolute certainty.

The various discoveries enumerated above furnish therefore all the materials necessary for the formation of an electric telegraph. Each inventor has, since such period, turned to this common stock of knowledge for the materials wherewith to build up his particular arrangements of telegraphic apparatus. One inventor has employed electricity produced by friction, another galvanic electricity, and a third magneto-electricity, and so on; and then each has used the apparatus most suited for the employment of the electricity so generated.

Having thus briefly noticed the discovery of the various component parts of an electric telegraph, it is proposed to proceed now to deal with the electric telegraph as a whole, and to notice as concisely as possible the particular arrangements recommended by various persons for the construction of a complete electric telegraph.

5
FIRST PRINCIPLES OF AN ELECTRIC TELEGRAPH

It is evident from the foregoing that if a wire were made to extend between London and Liverpool, and were insulated from the earth all the way between those two points, and if the *ends* of such wire were made to dip into the earth both at London and at Liverpool, that such wire would form a proper channel for the transmission of an electric current from either London to Liverpool or Liverpool to London.

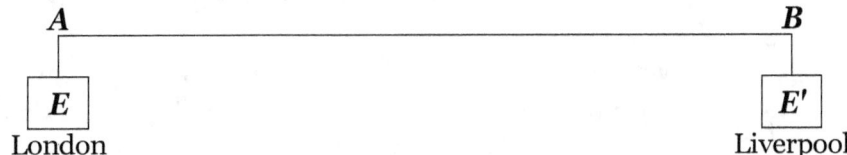

Fig. 2
Electric circuit between London and Liverpool

Suppose therefore in Figure 2 above, that *A B* represents such wire, and *E E'* the respective earth connections at London and Liverpool; then if the wire be severed in two at *A*, and the two severed ends be joined respectively with the two poles of a galvanic battery, as shown in Figure 3 below, it is evident that a positive current would flow from *P* through *A* to *E*, and a negative current from *N* through *B* to the earth at *E'*, and that it would return by the earth from *E'* to *E*, so as to complete its circuit and affect the

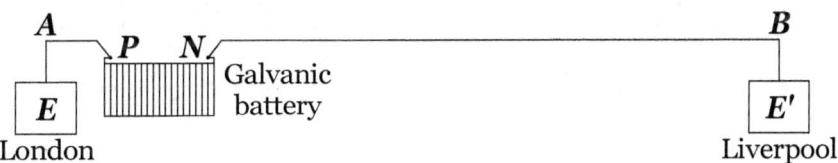

Fig. 3
Galvanic battery in circuit between London and Liverpool

apparatus in such circuit.

Again, if the positive pole **P** had been joined to the portion of wire extending to Liverpool, and the negative pole to the portion **A** at London, it is obvious that a positive current would pass to the earth at Liverpool and a negative one to the earth in London; and therefore that opposite electrical effects would have been produced on all instruments placed in the circuit of such wires.

Now, instead of turning the galvanic battery **P N** round, and thus connecting either the one pole or the other with the severed ends of the wire **A B**, it is evident that if the wire at such point of severance were joined to two keys in connection with the battery, and which keys were so arranged that the pressing of one down caused a positive current to progress to Liverpool and a negative one to London, and the pressing of the other down produced the contrary action, that by means of pressure upon either of two such keys either a positive or a negative current of electricity might be sent, as desired.

Again, if we suppose that when neither of such keys is pressed down, the ends of the severed wire at **A** (by very simple mechanical means) are allowed to *unite* metallically, the wire then becomes, electrically speaking, as it were unbroken or whole again. It might therefore, in the same manner, be severed at **B**, and the poles of another battery placed there might be attached to such severed ends at **B**, and by means of similar keys either a positive or negative current of electricity transmitted *from* Liverpool *to* London.

Such arrangement would enable a party at either London *or* Liverpool to cause currents of electricity to traverse the intervening wire, and similarly with respect to any number of intermediate

Fig. 4
Needle telegraph circuit between London and Liverpool

First Principles of an Electric Telegraph

stations placed in such circuit.

Now if at *both* London and Liverpool we put a coil of wire in the line circuit, as shown in Figure 4, and make such coil surround a magnetic needle *N S* fixed on a moveable axle, it is evident that when a current of positive electricity traverses the wire, the top of the needle will move in one direction, and when a negative current is sent, it will move in the other direction. If, then, such coil or needle be placed *vertically*, the top of the needle will move to the right by the one current and to the left by the other current.

This being clearly understood, it is evident that if one movement to the *right* represents the letter A, two successive movements to the *right* might represent B,—three movements C, and so on; and if one movement to the *left* represented M,—two movements to the *left* might represent N,—three movements O, and so on; and if one movement to the right, followed rapidly by one movement to the left, represented S,—two of such movements might represent T; and that by *commencing* with a movement to the *left* and *following* it by a movement to the *right*, another letter might be designated, and so on, with the various numbers of movements to the right or to the left; and with the combinations of one, two, three, or four of such movements in one direction with one or more movements in the other direction, all the letters of the alphabet could be designated.

Again, if instead of using a magnetic needle and coil, an electro-magnet had been employed with a moveable armature, it is evident that on sending a current the armature of such electro-magnet would be attracted to the electro-magnet, and if a pricker or marker were affixed to such armature and a piece of paper also were made to move near to such marker, every movement of the curvature caused by the electric current would be recorded on the moving paper by the point pressing against or going through the paper; and if the current were continued for a long period, the *length* of the scratch or mark would be proportionally great. By a combination therefore of dots and long and short marks, all the letters of the alphabet might be designated in this way; or, instead of an electro-magnet being employed, the current of electricity might be made to decompose chemical substances on a piece of moving paper and change the colour thereof. This would give the same means of enabling a person to read off the currents sent, and hence to understand the letters or signals intended to be transmitted. Similarly, also, with respect to all of the other enumerated means of testing the *presence* of an electric current transmitted through a conductor, whether it be made to move a magnetic needle, to mark paper, to produce sparks, to liberate mechanism, to remove screens, or to do any other pre-arranged work.

Let us now see how the above, as well as the many other plans of ascertaining the *presence* of an electric current in the *conducting wire*, have been employed by various parties for the purpose of telegraphic communication.

We will first describe those plans proposed prior to 1838.

6
TELEGRAPHS PRIOR TO 1837

It has been thought well, before describing the plans of telegraphs which have from 1838 to the present time formed the subject of patents in this kingdom, to notice briefly the principal features of the many telegraphs which preceded those for which patents were granted.

All these first telegraphs were freely given to the world by their respective inventors, and have furnished the materials employed by late patentees for their telegraphs. It is clearly very difficult, now that so many years have elapsed since these first telegraphs were invented, to fix the *precise* date at which the inventions were publicly known in *this* and *other* kingdoms.

The invention of the electric telegraph has, in different countries, been attributed to different individuals. Nothing, however, can be more incorrect than to attribute to any *one* man the invention of the electric telegraph, as so many eminent men have lent a helping hand in adapting the wonderful discoveries in electricity to the purpose of conveying intelligence. If to any single person the honour of having "invented the electric telegraph" is to be attributed, it surely ought to be either to the first person who proposed the employment of electricity for telegraphic purposes, or to the first person who did practically convey intelligence to a distant point by means of electricity. If so, then no *patentee* can claim the honour of inventing *The Electric Telegraph*.

But to proceed with a short summary of the peculiar features in the telegraphs invented prior to the grant of the first patent.

BRIEF SUMMARY OF TELEGRAPHS PRIOR TO THE YEAR 1837

- Lesarge, in 1774, employed 24 wires and a pith-ball electrometer.
- Lomond, in 1787, employed one wire and a pith-ball electro meter.

The Electric Telegraph–Its History and Progress

- Betancourt, in 1787, used one wire and a battery of Leyden jars.
- Reizen, in 1794, had 26 wires; the letters of the alphabet were cut out in pieces of tinfoil, and rendered visible by sparks of electricity.
- Cavallo, in 1795, used one wire; the number of sparks was made to designate the various signals, and the explosion of gas was used for an alarum.
- Salva, in 17 96.—The exact particulars of this telegraph are doubtful.

In all the above plans, high-tension electricity was to be employed.

- Soemmering's telegraph, of 1809 or 1811.—In this telegraph galvanic electricity was used, and as many wires were employed as there were letters or signals to be denoted. The letters were designated by the decomposition of water: an alarum was also added.
- Schwieger employed the principle of Soemmering's telegraph, but reduced the number of wires to two. He also proposed the printing of the letters.
- Coxe's telegraph, in 1810.—Coxe proposed the use both of the decomposition of water and also of metallic salts.
- Ronalds, in 1816. —In Ronalds telegraph high-tension electricity was employed. The wires used were laid underground as well as suspended in the air. A pith-ball electrometer, hung before a clock movement, enabled the letters on a dial to be read off. The sounding of an alarum by exploding gas, &c. was also added.
- Ampére, in 1820.—Ampére employed the magnetic needle, the coil of wire, and the galvanic battery, and proposed the use of as many wires as letters or signals to be indicated.
- Tribaoillet, in 1828.—Tribaoillet's telegraph required but one wire, and this was buried in the earth. A galvanic battery and a galvanoscope were employed.
- Schilling's telegraph, in 1832.—Schilling employed five magnetic needles and had also a mechanical alarum. In another telegraph of Schilling's one needle and one wire only were used.

Telegraphs Prior to 1837

- Gauss and Weber, 1833.—In the telegraph of Gauss and Weber one wire and one needle only were needed. The power employed was magneto-electricity.
- Taquin and Ettieyhausen, in 1836.—The particulars of the telegraph of these parties are at present uncertain.
- Steinheil's telegraph, 1837.—This telegraph required only one wire and one or two magnetic needles. The power used was magneto-electricity. Steinheil had a printing telegraph as well as a means of telegraphing by sounds produced by electric apparatus striking bells.
- Masson's telegraph, 1837 and 1838.—In this telegraph magneto-electricity was employed in conjunction with magnetic needles.
- Morse's telegraph, 1837.—Morse's telegraph was a printing or recording telegraph; it required only one wire, and galvanic electricity was used. An electro-magnet of iron was used for attracting an armature, to which was attached a pricker or pen to mark paper, which was made to pass underneath it.
- Vail's telegraph, 1837.—This was a telegraph for printing the letters of the alphabet. One wire only was used. Clockwork mechanism, regulated by pendulums, was also added.
- Davy's telegraph, 1837.—In this telegraph magnetic needles and coils of wire were used. The needles removed screens which previously rendered the letters invisible.
- Alexander's telegraph, 1837.—Thirty magnetic needles and thirty wires were required in this plan. Each needle removed a screen which obscured a letter painted behind it.

Previous to 1837 we have, therefore, no less than fifteen telegraphs, and in 1837 no less than six new arrangements of telegraphs, exclusive of the one of Messrs. Cooke and Wheatstone, which was patented in June, 1837.

Description of Telegraphs Prior to 1837

Lesarge's Telegraph.
The first electric telegraph of which we have any record is that of Lesarge. This telegraph was established at Geneva in 1774: it consisted of twenty-four wires, insulated from each other; each wire was in communication with a pith-ball electrometer placed at

the distant station. On sending an electric current over any one of these wires, the pith-balls which were attached to it diverged, and thus denoted the letter or symbol corresponding or affixed to that electrometer. In this way any one of twenty-four letters or signals could be instantly designated at pleasure.

LOMOND'S TELEGRAPH

The next telegraph we read of is that of Lomond. This telegraph consisted apparently of one wire, with an electrical machine and a pith-ball electrometer at each end thereof.

The signals were given by the divergence of the pith-balls, and as in this case there was but one wire with one electrometer at the terminus, the different signals were made by the number and variations of the divergences. It also appears that this telegraph was a reciprocal one, *i. e.* that the message could be sent from either terminus and received at the other, and similarly with respect to any intermediate stations.

A short account of this telegraph is to be found in Young's "Travels in France," at p. 979, vol. i. 4th edit. 1787. The telegraph is thus described:

> M. Lomond has made a remarkable discovery in electricity. You write two or three words on a paper; he takes it into a room, and turns a machine enclosed in a cylindrical case, at the top of which is an electrometer, a small fine pith-ball; a wire connects with a similar cylinder and electrometer in a distant apartment; and his wife, by remarking the corresponding motions of the ball, writes down the words they indicate, from which it appears that he has formed an alphabet of motions. As the length of the wire makes no difference in the effect, a correspondence might be carried on at any distance, within or without a besieged town for instance, or for objects much more worthy of attention and a thousand times more harmless.

BETANCOURT'S TELEGRAPH

In the same year in which the above description was published, Betancourt, by means of a wire extending between Aranjuez and Madrid, a distance of twenty-six miles, transmitted signals by passing discharges of electricity from a battery of Leyden jars.

The full particulars of this kind of telegraph the author has not been enabled to obtain, but enough is known to show that wires of considerable length were employed, even at that time, to enable parties to transmit telegraphic signals by means of electricity to places at a considerable distance.

REIZEN'S TELEGRAPH

The first notice we find of Reizen's telegraph is inserted in the "Magazine de Voight," in 1794. His plan was as follows:

As many wires were to be insulated and laid in glass tubes to a distant station as there were letters of the alphabet or different signals to be designated; each wire was to communicate with strips of metal placed upon a square of glass; the strips of metal were to be formed in the shape of letters, and instead of being of continuous metal, several breaks were to intervene, so that in the passage of the electricity a bright flash would be seen at every break. Thus, when these breaks in a letter were many, and a current was passed through, the letter appeared illuminated from one end to the other.

A wire of the telegraph being put into metallic communication with the commencement of each letter or symbol, when a discharge was sent from the electrical machine or from a Leyden jar over that wire, the breaks forming the letter became illuminated, and hence the letter was made visible. In this way letter after letter might be exhibited, and hence a correspondence carried on thereby.

Figure 5, showing the tinfoil with the breaks therein, will explain Reizen's mode of communicating intelligence by means of electricity.

CAVALLO'S TELEGRAPH

Mons. Cavallo. in his "Traité de Electricité," published in 1795, gives a description of his mode of conveying intelligence by means of electricity.

His plan was first to call attention by means of the explosion of a detonating substance,—such as gunpowder, a mixture of hydrogen and oxygen, or phosphorus,—and then to convey the intelligence by transmitting a number of sparks from the Leyden jar before a slight pause. The letters and signals were made by

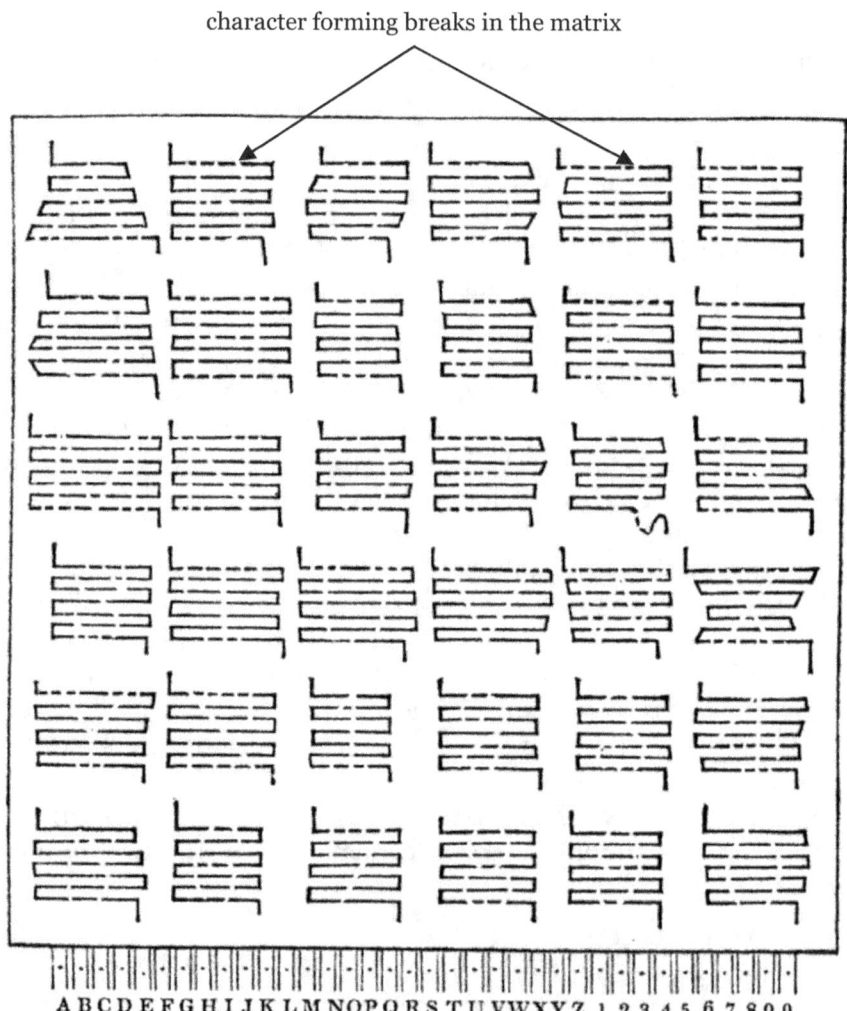

Fig. 5
Reizen telegraph tinfoil matrix, with breaks that shape letters and numerals

counting the number of sparks so sent before a pause.

SALVA'S TELEGRAPH

An account of Dr. Salva's telegraph was given to the Academy of Sciences at Paris in 1796; and in November of the same year, the "Gazette de Madrid" contained an article thereon. The article runs thus:

> The "Prince de la Paix," having learned that M. de F. Salva had read at the Academy of Sciences a memorial of an application of electricity to the telegraph, and presented at the same time an electric telegraph of his invention, wished to examine it; and, charmed with its promptitude and the facility of its operations, he showed it afterwards to the King and to the Court, when it performed equally well.
>
> After this experiment, the young Don Antonio wished to obtain a more perfect telegraph, and undertook to calculate the force of the electricity required to work a telegraph at different distances under land or water.
>
> Some useful experiments have been made, which we shall speak of hereafter.

The "Magazine de Voight," in reference to these experiments, announced two years afterwards that Don Antonio constructed a telegraph upon a very grand scale and to a very great extent. It also states that the same young Prince was informed at night, by means of this telegraph, of news that highly interested him.

It appears therefore that this telegraph was no chimera, but one that was capable of sending a very considerable amount of information, or otherwise it could not have transmitted a piece of news.

SOEMMERING'S TELEGRAPH

Soemmering's telegraph appears to have been invented in 1809, though some parties give to it the date of 1811.

This telegraph was the first in which the pile or first form of the galvanic battery was used.

Soemmering's plan was as follows: Insulated wires were to be laid to the distant station; these wires were to terminate at the receiving station in gold points, which were to dip into a glass vessel containing acidulated water; each gold point represented a letter of the alphabet.

At the transmitting station the wire terminated with mechanical arrangements, so that the end of any one wire could be brought into metallic contact with one end of the pile or battery, and any other wire with the other end of the battery. In this way an electric current could be sent down any one of the wires, and return by any other. (This *principle* was afterwards used by Professor Wheatstone, in his patent of 1837, which required five or six wires for each telegraph.)

When a current was so sent from the galvanic pile over any two wires, the gold points connected with those two wires at the distant station gave off bubbles of oxygen and hydrogen gases; and the two letters corresponding therewith were thus denoted. If another wire representing 0 (or nothing) were added, it is evident that either two letters, or only one, could be denoted at pleasure.

From the experiments made by Soemmering, and the instantaneous appearance of the gas when the battery was thrown into the circuit, he concluded that the passage of the power was instantaneous. He also found that the addition of 2,000 feet of wire in the length of his circuit produced little or no sensible additional resistance, and that for nearly 3,000 feet of wire, the decomposition of the water, and the appearance of the gas at the distant station, commenced instantaneously with the sending of the current.

Soemmering used in his apparatus 35 wires, 25 of which were for the letters of the German alphabet, the remainder for the 9 numerals, and one for 0. The wires were insulated by means of silk. This, it must be observed, is the first galvanic telegraph, and it is a telegraph by which, by a single movement, any one or any two letters of the alphabet could instantly be denoted.

No mention was made at the time of the means to be employed for first calling the attention of the attendant at the distant station, but Soemmering afterwards proposed to liberate a wound-up alarum by means of the evolution of gas,—a plan which appears to have been patented afterwards by Messrs. Cooke and Wheatstone in 1837.

The following is Soemmering's own description of his telegraph:

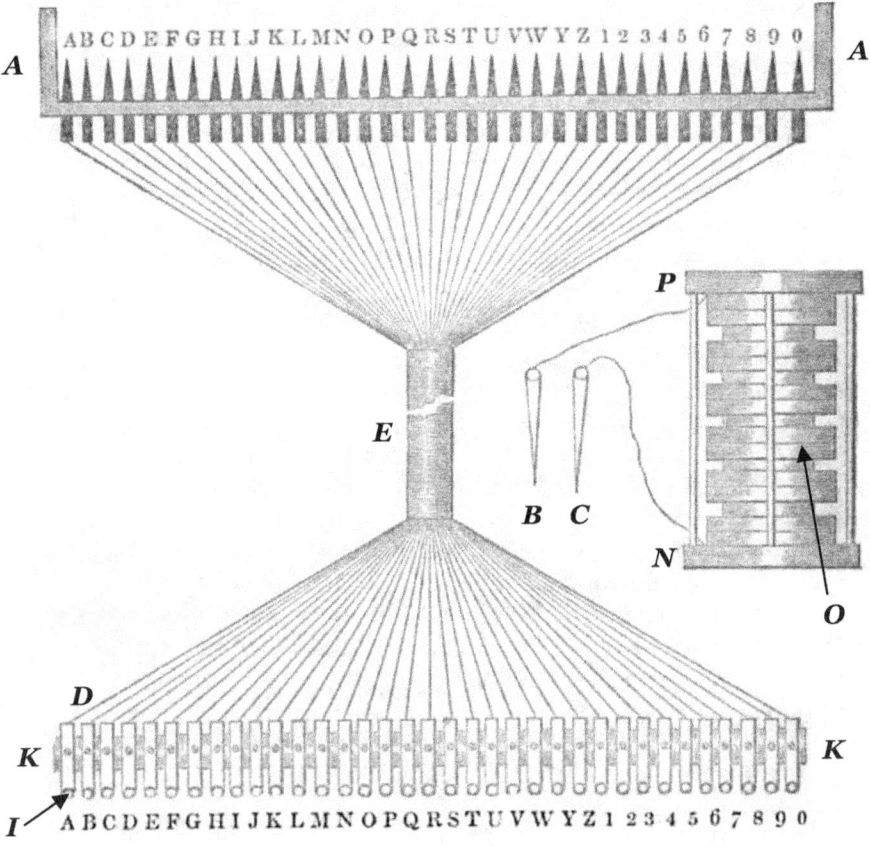

Fig. 6
Soemmering's telegraph, which used the decomposition of water by electricity to designate letters and numerals

"My telegraph was constructed and used in the following manner: In the bottom of a glass reservoir, shown in Figure 6, of which *A A* is a sectional view, are 35 golden points or pins, passing up through the bottom of the glass reservoir, marked A, B, C, &c., and thus bearing the 25 letters of the German alphabet and the ten numerals.

The 35 points are each connected with an extended copper wire, soldered to them, and, extending through the tube *E* to the distant station *D*, are there soldered to the 35 brass plates, upon the wooden bar, *K K*. Through the front end of each of the plates there is a small hole, *I*, for the reception of two brass pins, *B* and *C*; one of which is on the end of the wire connecting the positive pole, and the other the negative pole, of the voltaic column, *O*. Each of the 35 plates is arranged upon a support of wood, *K K*, to correspond with the arrangement of the 35 points at the reservoir, and is lettered accordingly. When thus arranged, the two pins from the voltaic column are held one in each hand, and the two plates being selected, the pins are then put into their holes and the communication is established.

Gas is evolved at the two distant corresponding points in an instant; for example, K and T.

The peg on the hydrogen pole evolves hydrogen gas, and that on the oxygen pole, oxygen gas. In this way every letter and numeral may be indicated at the pleasure of the operator. Should the following rules be observed, it will enable the operator to communicate as much if not more than can be done by the *common telegraph*.

- *First Rule.* As the hydrogen gas evolved is greater in quantity than the oxygen, therefore those letters which the former gas represents are more easily distinguished than those of the latter, and must be so noted. For example, in the words containing *ak, ad, em, ie,* we indicate the first letters *a, a, e, i,* by the hydrogen gas; *k, d, m, e,* on the other hand, by the oxygen gas.
- *Second Rule.* To telegraph two letters of the same name, we must use a unit, unless they are separated by the syllable. For example, the word *anna* may be telegraphed without the unit, as the syllable *an* is first indicated and then *na*. The word *nanni*, on the contrary, cannot be telegraphed without

the use of the unit, because *na* is first telegraphed, and then comes *nn*, which cannot be indicated in the same vessel. It would, however, be possible to telegraph even three or more letters at the same time by increasing the number of wires from 25 to 50, but which would very much augment the cost of construction and the care of attendance.

- *Third Rule.* To indicate the conclusion of a word, the unit 1 must be used. Therefore it is used with the last single letter of a word, being made to follow the ending letter. It must also be prefixed to the letter commencing a word when that letter follows a word of *two letters only*. For example: *sie lebt* must be represented *si*, *e*1, le, bt; that is, the unit 1 must be placed after the first *e*. On the contrary *er lebt*, must be represented *er*, 1*l*, *eb*, *t*1; that is, the unit 1 is placed before the *l*. Instead of using the unit, another signal may be introduced, say the cross + to indicate the separation of syllables.

Suppose now the decomposing table is situated in one city, and the pin arrangement in another, connected with each other by 35 continuous wires, extended from city to city. Then the operator, with his voltaic column and pin arrangement at one station, may communicate intelligence to the observer of the gas at the decomposing table of the other station.

The metallic plates with which the extended wires are connected have conical-shaped holes in their ends; and the pins attached to the two wires of the voltaic column are likewise of a conical shape, so that when they are put in the holes, there may be a close fit, preventing oxidation and producing a certain connection. It is well known that slight oxidation of the parts in contact will interrupt the communication. The pin arrangement might be so contrived as to use permanent keys, which for the 35 plates or rods would require 70 pins. The first key might be for hydrogen A; the third key for hydrogen B ; the fourth key for oxygen B, and so on.

The preparation and management of the voltaic column is so well known that little need be said, except that it should be of that durability as to last more than a month. It should not be of very broad surfaces, as I have proved that six of my usual plates (each one consisting of a Brabant dollar, felt, and a disc of zinc, weighing 52 grains) would evolve more gas than five plates of the great battery of our Academy.†

As to the cost of construction, this model, which I have had the honour to exhibit to the Royal Academy, cost 30 florins. One line, consisting of 35 wires, laid in glass or earthen pipes, each wire insulated with silk, making each wire 22,827 Parisian feet, or a German mile, or a single wire of 788,885 feet in length, might be made for less than 2,000 florins, as appears from the cost of my short one."

SCHWIEGER'S TELEGRAPH

Schwieger proposed, that instead of using so many wires as required by Soemmering, the number should be greatly reduced. His preposition was that *two* galvanic piles should be used, the one considerably more powerful than the other; so that at one time the weaker one might be used, at another the stronger one, and at another both combined.

By this arrangement the amount of gas given off in a given period of time at the distant station would be varied. When a small quantity of gas was being evolved, one letter might be denoted, and when a larger volume was produced, a different letter, and so on. And again, if the *periods of time* during which this evolution of gas were varied also, other and different letters might be denoted. In this way, and by combinations of these different primary results, did Schwieger propose to reduce the number of wires down to two, and yet to be able to denote every letter of the alphabet.

Schwieger also proposed methods of permanently registering the letters denoted; this was to be done by means of paper smeared with lamp-black and other substances,—a plan long afterwards employed by Professor Wheatstone and partly patented by him.

COXE'S TELEGRAPH

In Thompson's "Annals of Electricity," in 1810, Professor Coxe, of Philadelphia, alludes to certain plans of telegraphing by means of the galvanic pile. He appears to have had two plans—the one being by the decomposition of water at distant stations, and the other by the decomposition of metallic salts.

---

† **Author's Note:** Academy of Sciences at Munich.

Thus we see that as each successive discovery in the *effects* produced by electricity became known, ingenious men in all parts of the world turned their attention almost immediately to the application of those very discoveries to the art of telegraphing.

RONALDS' TELEGRAPH

In 1816 and the following years Mr. Ronalds, of Hammersmith, devoted much time to the investigation of the electric telegraph. He erected eight miles of insulated wire on his lawn; he also buried a considerable length of insulated wire in the earth. The wires in the air were insulated by silk and dry wood, and those in the earth by enclosing the wire in glass tubes, surrounded by a wooden trough filled with pitch.

He employed the ordinary electric machine and the pith-ball electrometer in the following manner. He placed two clocks at two stations; these clocks had upon the second-hand arbour a dial with twenty letters on it; a screen was placed in front of each of these dials, and an orifice was cut in each screen so that one letter only at a time could be seen on the revolving dial. These clocks were made to go isochronously, and as the dials moved round, the same letter always appeared through the orifices of each of these screens. The pith-ball electrometers were hung in front of the dials. It is evident, therefore, that if these pith-balls could be made to move at the same instant of time, that a person at the transmitting station, by causing such motion in both those electrometers, would be able to inform the attendant at the distant or receiving station what letters to note down as they appeared before him in succession on the dial of the clock.

This was accomplished in the following manner. The transmitter caused a current of electricity to be constantly operating upon the electrometers, so as to separate the balls of those electrometers, except only when it was required to denote a letter, and then he discharged the electricity from the wire, and instantly both balls collapsed. The distant observer was thereby informed to note down the letter then visible. In this way letter after letter could be denoted, words spelt, and intelligence of any kind transmitted. All that was absolutely required for this form of telegraph

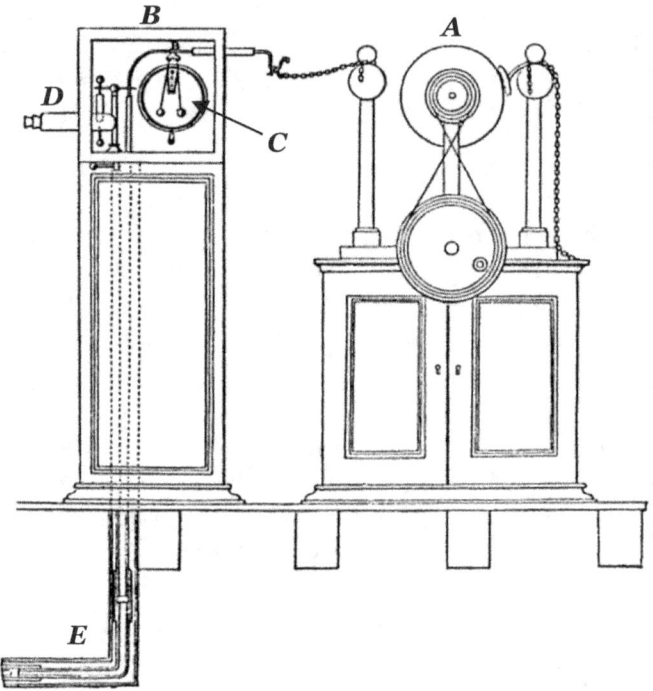

Fig. 7
Ronalds' pith-ball and screen telegraph on the left,
electricity generator on the right

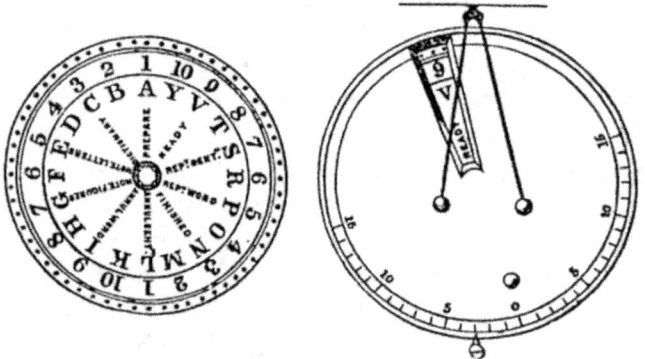

Fig. 8
Enlargement of the movable dial on the left, the screen
on the right with the cut-out revealing the movable dial,
and the pith-ball electrometer

was, that the clocks should go isochronously *during the time* that the intelligence was being transmitted, for it was easy enough by a preconcerted arrangement between the parties, and upon a given signal, for each party to start their clocks at the same letter, and thus if the clocks went together during the transmission of the intelligence, the proper letters would appear simultaneously, until the communication was finished. The attention of the distant observer was called by the explosion of gas by means of electricity from a Leyden jar.

Mr. Ronalds, in 1823, published a full description of this telegraph, in a work entitled "Descriptions of an electrical telegraph, and of some other electrical apparatus." Figure 7 shows Ronalds' Telegraph.

A is an electrical machine; *B*, the pith-ball electrometer; *C*, the screen hiding the letters on the dial behind it; *D*, the gas alarum; *E*, the tube conveying the wires.

Figure 8 is an enlarged drawing of the moveable dial hidden by the screen, and the screen with cut-out and pith-ball electrometer.

Mr. Ronalds enters on the subject of the comparative merits of wires suspended in the air and wires buried in the earth, and arrives at the conclusion that subterranean wires are much to be preferred, although many persons were found to object to that plan.

He says:

> "The liability of the subterranean part of the apparatus to be injured by an enemy or by mischievously disposed persons has been vehemently objected to—more vehemently than rationally, I presume to hope (as is not infrequently the case on these as on many other sorts of occasions). If an enemy had occupation of all the roads which covered the wires, he could, undoubtedly, disconcert my electric signs without difficulty; but would those now in use escape? And this case relates only to invasions and civil wars; therefore let us have smokers enough to prevent invasions, and kings that love their subjects enough to prevent civil wars.
>
> To protect the apparatus from mischievously disposed persons, let the tubes be buried six feet below the surface of the middle of the high roads, and let each tube take a different

route to arrive at the same place. Could any number of rogues then open trenches six feet deep, in two or more different public high roads or streets, and get through two or more strong cast-iron troughs, in less space of time than forty minutes? For we shall presently see that they would be detected before the expiration of that time. *If they could* render their difficulties greater by cutting the trench deeper, and should they still succeed in breaking the communication by these means, hang them if you catch them, damn them if you cannot, and mend it immediately in both cases."

Fortunately, however, there is now no need of either hanging or damning the rascals, for the law makes a wilful damage done to the electric telegraph a misdemeanour, and punishes accordingly.

AMPÉRE'S TELEGRAPH

Immediately after the brilliant discovery of Oersted in 1809, (viz. that a magnetic needle always tends to place itself at right angles to an adjoining wire through which an electric current from a galvanic battery is passing,) Ampére read a paper before the Academy of Sciences, at Paris, in 1820, on a plan for an electric telegraph, based upon this discovery of Oersted.

Its *principle of action* was the one subsequently used by Professor Wheatstone in all his needle telegraphs, and therefore to Ampére, who first proposed the use and combination of the magnetic needle—the coil of wire—and the galvanic battery, is due the credit of being the first person to publish to the world the perfect practicability of an electric telegraph constructed *with magnetic needles surrounded by coils of wire*, and moved by electricity generated in the galvanic battery.

Ampére proposed to use as many needles as there were letters or symbols required to be denoted. Wheatstone, by combining the apparatus of Ampére in a peculiar way, obtained twenty letters by the use of only five wires, as will be explained hereafter, and, therefore, greatly improved upon Ampére's plan.

The plan of Ampére was as follows: He proposed to have at every station from which intelligence was to be sent, a galvanic battery, with all necessary keys for putting the battery in communication with the wires, and to have at the points where intelligence was to be received as many magnetic needles as there were letters required to be denoted. Each letter was placed upon a different needle, and the needles were surrounded with coils of wire in metallic communication with the wires extending between the stations. It is evident, therefore, that upon the transmission of a current of electricity through any one of those coils the needle would move, and with it the letter, and thus letter after letter would be denoted.

Here, then, we have the first good needle telegraph, and one all but perfect in its parts, with the exception only of the *great number of wires* required to be employed.

Ampére, it is true, might not have given all the minor details as to the keys, &c., nor was it in any way needed at the time, inasmuch as anyone who had ever used a battery and tried experiments therewith, would at once have known how to make suitable keys in a dozen different ways, and how to convert a receiving station into a transmitting station whenever required to do so. Much misapprehension on this point seems to have prevailed of late in the minds of certain scientific men in this kingdom; but the circumstances under which those opinions have been expressed are so peculiar, that every allowance must be made for the parties.

So well understood does the magnetic needle telegraph seem to have been in 1827, that Dr. Green, who wrote at that time, says:

> "In the very early stage of electro-magnetic experiment it had been suggested that an instantaneous telegraph might be constructed by means of conjunctive wires and magnetic needles. The details of this contrivance are so obvious, and the principles on which it is founded are so well understood, that there was only one question which could render the result doubtful. This was, whether, by lengthening the conjunctive wires, there would be any diminution in the electrical effect upon the needle."

It was evident, therefore, at this time that nothing was wanting but direct experiment to test this, then doubtful, point, as to how far galvanic electricity would travel through a wire. Professor

Barlow's opinion was, that the force of the current would be so diminished by the length of the circuit, that a galvano-electric telegraph would, for long distances, be impracticable. Other scientific men differed with Professor Barlow in this opinion. To Professor Wheatstone is due the credit of having *practically* solved the question, in ascertaining the relative resistances of the parts of the circuit of an electric telegraph, and also of adjusting the size and length of wire required for the magnets to be employed. Professor Ohm founded the mathematical expression for the law of the resistances to the passage of *all* electric currents from the galvanic battery. These resistances were expressed not only in terms of the line-wire, but in terms of the size and distance of the battery-plates from each other, and also of the resistance of the fluid in the battery itself. Professor Wheatstone, therefore, had only to apply this general law to the peculiarities of the electric telegraph, and the problem would become at once satisfactorily solved, and it would show that to the extent of at least 1000 or 2000 miles the use of galvanic electricity for the purposes of the electric telegraph was perfectly practicable.

TRIBOAILLET'S TELEGRAPH

Mr. Triboaillet, in 1828, proposed the following arrangement for an electric telegraph:

A single wire only was to be used. The wire was to be covered with shellac, then wrapped with silk, and afterwards covered with resin. This insulated wire was then to be buried in the earth, inside glass tubes, the joints being carefully luted up and made water-tight. The electricity was to be generated by a powerful battery, and to act through the insulated wire on a delicate electroscope at the distant station.

Mr. Triboaillet prepared no particular form of code for his telegraph, but he left it to each telegraphist to form his own alphabet, on the principle of making the number of the motions to express the various letters or symbols desired to be denoted, as is now done with respect to the needle telegraph at present in use in England.

Schilling's Telegraph

M. Le Baron de Schilling appears to have invented two kinds of telegraph; one with five magnetic needles, and another with only one needle. The first had five needles, and was constructed at St. Petersburg in 1832.

By the single deflection of each of these five needles to the right or to the left, ten primary signals were obtained, and by means of a code or dictionary the combination of a few of such signals was made to express whole words or sentences. Schilling also invented an alarum. The motion of one of his magnetic needles allowed a weight to fall, and by the momentum, produced by such fall, to cause an alarum to sound.

Another of Schilling's plans, and of apparently later date, was to use only *one* magnetic needle, and by counting the number of such motions of that needle to the right and left, to designate the letters of the alphabet thereby.

The telegraphs of Schilling were exhibited before the Emperor Alexander, as well as afterwards before the Emperor Nicholas, and were highly approved of by both.

Gauss and Weber's Telegraph

In 1833 a telegraph was invented by Gauss and Weber at Gottingen. This consisted of a magnetic needle surrounded by a coil of wire, the needle being moved by the agency of electricity developed by the magneto-machine. The electricity generated was not sent simply in intermittent currents, but by means of mechanical contrivances, then well known, a constant current was produced so as to cause the deflection of the magnetic needle to continue for any desired period. The signals were to be made by the number of deflections to the right and left. When a total of five motions was made for each signal, the number of different signals transmitted would amount to more than all the letters of the alphabet, as well as all the numerals, and many spare signals for special objects would thus be produced.

This telegraph was constructed at Gottingen between the Observatory and the Cabinet de Physiques (a distance of a mile and a quarter). The earth appears to have been used as part of

The Electric Telegraph–Its History and Progress

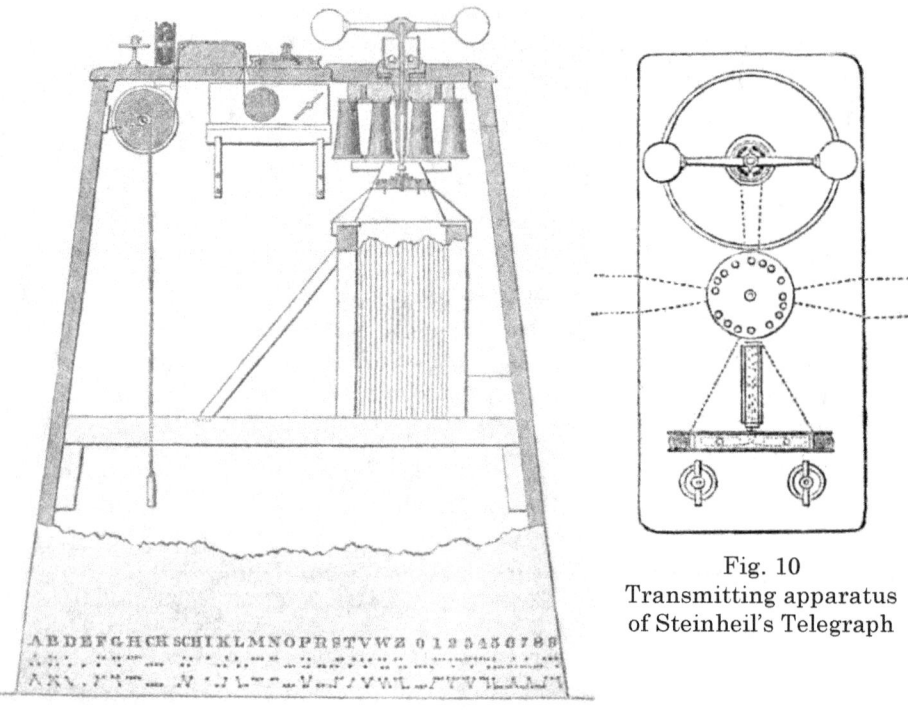

Fig. 10
Transmitting apparatus of Steinheil's Telegraph

Fig. 9
Steinheil's Telegraph

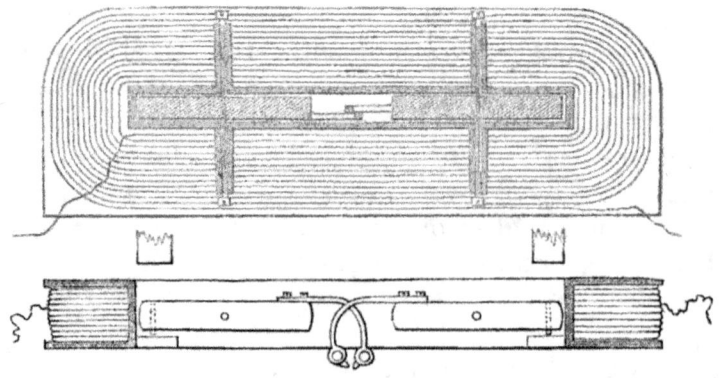

Fig. 11
Coils and magnets of Steinheil's Telegraph

the circuit.

TAQUIN AND ETTIEYHAUSEN'S TELEGRAPH

It appears that Messrs. Taquin and Ettieyhausen established a line of telegraph across two streets in Vienna in 1836. The wires were partly suspended in the air and partly buried in the earth.

The journal that affords the above information does not give any other particulars, nor does it state the kind of instruments employed.

STEINHEIL'S TELEGRAPH

This telegraph, in point of time, precedes the first patented telegraph in England. It was also a perfect arrangement. Dr. Steinheil could either telegraph by sound or by the making of permanent marks on paper: he employed both these different processes.

His telegraph consisted of *one* wire, and of one or two magnetic needles as desired. The needles, as in former plans, were surrounded by coils of wire, and each could be made to move to the right or left by electricity generated from the magneto-electric machine. When it was desired to telegraph by sound, he made the needles strike against either of two bells,— the one needle striking one bell, and the other needle striking another, differently toned. When he required to permanently record the intelligence, these needles were furnished with small tubes holding ink, and by their motions dots were made on paper properly moved in front of them by wound-up mechanism; one needle making dots in one line, and the other needle making dots in a line underneath the former.

Twelve miles of wire were erected, and intermediate as well as terminal stations employed. A portion of the wire was covered with zinc, and the ends of the wire at each distant terminus were joined to plates of metal buried in the earth, so that the earth formed one-half of the circuit.

On the whole, Steinheil's telegraph is a very perfect one, and may well put to shame many of the plans afterwards patented in this kingdom.

He made his signals by a maximum of *four* dots. He used galvanized iron wire. He employed but one wire for his telegraph. He used the earth circuit; and he carried wire both underground and in the air.

One thing only seemed needed to make this telegraph perfect, and that was a means of employing secondary power for printing or marking the paper,—this has now been accomplished by the author.

Steinheil's telegraph was in practical operation in July, 1837, was twelve miles long, and had three stations in the circuit. Figures 9, 10 and 11 illustrate Steinheil's Telegraph.

During the year 1837 many telegraphs were invented. This is also the year in which the first patent was taken out for an electric telegraph in England. It is thought better, however, to finish the description of all those telegraphs that were not patented in England in 1837, before commencing with those patented.

MASSON'S TELEGRAPH

At Caen, in 1837, M. Masson erected a line of telegraph about a mile and a quarter long.

The power he employed was electricity, developed from the magneto-machine, which was made to operate on magnetic needles at the respective termini.

In 1838, M. Masson in conjunction with M. Breguet tried further experiments on a line of railway.

MORSE'S TELEGRAPH.

Professor Morse has stated that he invented his telegraph in 1832: it does not appear, however, that any telegraph was actually constructed, nor the thoughts of his brain put in practice until 1837. Morse himself, in a letter to the Secretary of the Treasury of the United States, dated September 27, 1837, says:

> "About five years ago, on my voyage home from Europe, the electrical experiment of Franklin upon a wire some four miles

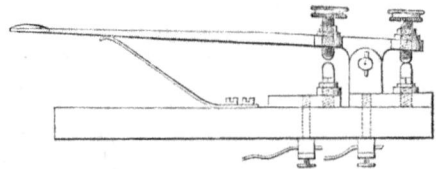

Fig.. 12
Morse key for sending currents of electricity

in length was casually recalled to my mind in a conversation with one of the passengers, in which experiment it was ascertained that the electricity travelled through the whole circuit in a time not appreciable but apparently instantaneous.

It immediately occurred to me that if the presence of electricity *could be made **visible** in any part of this circuit, it would not be difficult to construct a system of signs* by which intelligence *could be instantaneously transmitted.*

The thought thus conceived took strong hold of my mind in the leisure which the voyage afforded, and I planned a system of signs and an apparatus to carry it into effect. I cast a species of type which I devised for this purpose, the first week after my arrival home; and although the rest of the machinery was planned, yet from the pressure of unavoidable duties I was compelled to postpone my experiments, and was not able to test the whole plan until within a few weeks. The result has

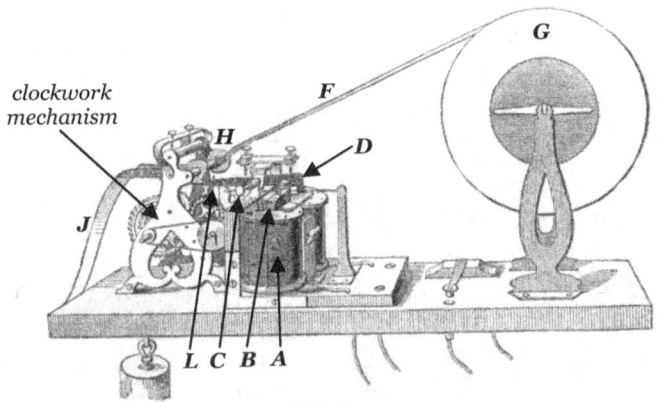

Fig. 13
Morse telegraph recording instrument

Fig. 14
Morse American alphabet

realized my most sanguine expectations.

As I have contracted with Mr. Alfred Vail to have a complete apparatus made to demonstrate at Washington, by the 1st of January, 1838, the practicability and superiority of my mode of telegraphic communication by means of electro-magnetism, (an apparatus which I hope to have the pleasure of exhibiting to you,) I will confine this communication to a statement of its peculiar advantages..."

This letter appears to be an answer to a circular dated March 10, 1837, sent to certain collectors of the Customs, &c., with reference to telegraphic communication, and desiring a reply by the 1st of October, 1837.

Now, whether the date of Morse's telegraph is to commence from the period when the thought first passed across his mind, or from the time when the first telegraph was made and signals actually produced by him, I must leave it with others to decide.

Much controversy has taken place already on this point, and no doubt many will, even now, with all the facts before them, come to different conclusions thereon.

The peculiarities of Morse's telegraph, when made, were the use of one wire, and that wire either to be placed underground or in the air. A galvanic battery at the transmitting station was to furnish the power, and an electro-magnet of iron at the receiving station was to record the presence or passage of the power.

The armature of this electro-magnet was to have attached to it a pen with ink in, or a pencil, for the purpose of marking paper, which was to pass uniformly along in front of the pen. The pencil or pen was afterwards abandoned for the use of a steel pricker.

The first symbols used were characters like a V; afterwards, when the pricker was used, it made small holes in the paper, or formed long scratches on it, accordingly as the current of electricity was kept on for a short or long period. The combination of dots and long strokes thus formed his alphabet. This telegraph has been most extensively used in America, and is very simple both in construction and use.

The first experiment was made over half a mile, on the 2nd October, 1837.

Telegraphs Prior to 1837

Figure. 12 shows the lever key usually employed for transmitting the currents of electricity. This is too simple to need description.

Figure. 13 is the recording or receiving instrument. *A* is the electro-magnet; *B*, the armature attached to the lever *D*, working on the centre *C*; *L* is the steel pricker; *F* is the long strip of paper coming off the drum *G*, and passing under the roller *H* ; *J* is the paper after it has been marked by the pricker.

The paper is kept continually moving under the roller *H* by means of clockwork mechanism, as shown.

When an electric current passes round the electro-magnet *A*, the armature *B* is drawn down, and the pricker *L* forced into the strip of paper.

If the current be held on for a long period, a long mark or cut will be made in the paper,—if for a short time, a dot only is produced.

Figure 14 represents Professor Morse's alphabet, composed, as it will be observed, of long and short marks.

VAIL'S TELEGRAPH

From Vail's work on Telegraphs it appears that in September, 1837, while he was engaged in making a telegraphic instrument for Professor Morse, for the purpose of exhibiting its actions to a Committee of the Congress, he invented a telegraph for printing the letters of the alphabet.

The arrangement appears to have been as follows: At the receiving station an ordinary clockwork apparatus was employed with an escapement action; to the axle of the seconds wheel a type wheel was affixed; as this seconds wheel moved step by step by means of a pendulum, the type wheel was moved forward letter by letter; every letter was thus brought successively under a type connected with an electro-magnet in metallic connection with the line-wire. This electro-magnet was so arranged that when a current of electricity passed through it, the paper was pressed against the type wheel, and an impression made of the letter then present.

At the transmitting station a similar instrument was employed but in the type wheel, holes were drilled opposite each letter, in order to insert therein a small pin. As the type wheel revolved by the action of the clockwork apparatus, a pin so inserted in any of the holes in the type wheel came in contact with a projecting piece of brass, and a current was thus caused to pass along the line-wire, and the letter opposite to that pin printed.

The pin was then suddenly removed and placed in a hole corresponding with the next letter to be printed; and then that letter was printed, and so on to the end of the message.

This telegraph is a very complicated one: the limited space allowed prevents a fuller account being given: an extended description will be found in the account of the American telegraph as published by Mr. Vail.

ALEXANDER'S TELEGRAPH

In 1837, Mr. Alexander, of Edinburgh, constructed an electric telegraph on the following plan:

He employed 30 wires and 30 magnetic needles. At the end of each needle was fixed a screen covering a letter behind it: on the transmission of a current of electricity from a galvanic battery over any wire, the corresponding needle was moved to one side, and the desired letter exposed to view. The letters were painted on a vertical dial, and the needles arranged over them; 30 keys were used with the letters of the alphabet on them. On pressing down any one of these keys a current of electricity was made to traverse its wire and to act on the needle belonging to that wire. By confining the motion of each needle to one direction only, no oscillation or vibration of the screen ensued.

This telegraph was exhibited at work at the Society of Arts in Edinburgh, in 1837, and an account of it appeared in some of the Scotch papers, as well as in the Mechanic's Magazine for November, 1837.

The original instrument was lately shown at the Great Exhibition in Hyde Park. Figure 15 is a drawing of it:

A is a voltaic battery; *B* a trough filled with mercury; *C*, a wire connecting the zinc plate in the battery with the trough of mercury; *D*, the return wire connected with the copper plate of

Telegraphs Prior to 1837

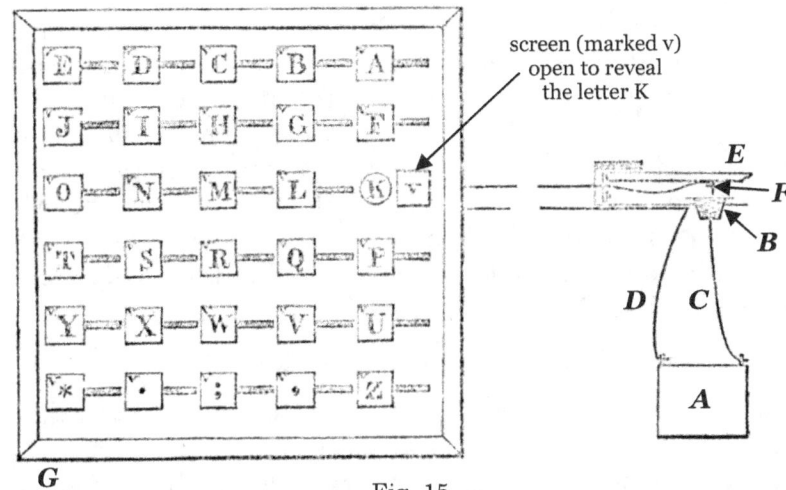

Fig. 15
Alexander's indicating telegraph

the battery; ***E***, a key to be pressed down by the finger of the operator, like the key of a pianoforte; ***F*** is a pendant wire which dips into the mercury when the key is depressed, and completes the circuit formed by the wires ***C*** and ***D***, extending from one terminus of the telegraph to the other.

G is the distant dial upon which the letters and punctuation characters of the alphabet and are marked. The letters are not seen when the magnetic needles—poised horizontally in free space behind the dial—are in their natural position of north and south, but are hidden by screens or veils marked v, attached to each of their north poles. But when the circuit is completed by the depression of the key ***E***, the corresponding magnetic needle is deflected to the west and exposes, as at K, the letter previously concealed. Thirty copper wires and a return wire extend from the keys to the magnetic needles.

A metallic rod may be advantageously substituted for the trough of mercury below the keys.

DAVY'S TELEGRAPH

The arrangement of Davy's telegraph was as follows: Small screens were attached to magnetic needles surrounded by coils of wire; behind the screens were the letters of the alphabet painted upon ground glass; a lamp was placed behind the instrument so as to illuminate the letters which were delineated on the glass.

On the removal of a screen to the right, one letter was brought into view, and on its removal to the left, another letter, and so on. Twelve keys were used. A galvanic battery was employed to cause the needles to move, and with them the screens. The exact number of wires does not appear.

An account of this telegraph appeared in the London Mechanic's Magazine for 1837.

History of Telegraph Patents

We come now to the period when every improvement in the electric telegraph in England was made the subject of a patent, but it would be impossible to give an exact description of each of the parts of the patents of the various patentees in the space allowed.

It would also be utterly impossible to make many parts of the inventions intelligible to the reader without having engravings of all the drawings to accompany the specifications; and to give a full description of the many patents, with engravings of the various drawings accompanying the respective specifications, would entail an enormous expense.

To supply the reader with an approximate idea as to what the cost would be, it will only be necessary to state, that to obtain a written copy of the specification of almost any of these patents would of itself cost from £10 to £40—and mere office copies of the specification of two patentees would cost £169. 98s.

There are now no less than about 40 patents taken out in England in connexion with electric telegraphs, so that the mere cost of obtaining copies of the specifications of the English patents, and irrespective entirely of the cost of printing the same, and having engravings made for them, would totally preclude the possibility of giving each of them *in extenso*.

It is to be hoped that the laws respecting patents will not remain long in their present unsatisfactory state.

Nothing can be more absurd than the present rules with respect to the specifications of patents. Every individual in this kingdom is as much bound by the specifications of a patent as by an Act of

Parliament. No one may either *make, use, exercise,* or *vend* any patented article, unless permitted to do so by the patentee. Hundreds of patents are taken out annually, and yet a man living anywhere out of London, unless he comes up to town and searches in the Enrolment Offices of the Courts of Chancery,—a process which might detain him many days at a large expense,—cannot know what he is forbidden by law either to *make, use,* or *sell.*

Again, no classified index is kept at the Enrolment Office, so that unless a person employs a patent agent, or some good fortune attend him, he may search for days and weeks in a variety of books and rolls of parchment, without knowing whether he is or is not forbidden to make an article which he has just invented, but which may, for aught he can learn to the contrary, have been, during the previous fourteen years, patented by someone else.

Until a proper index is made of all patents, and published periodically, and sold at rates similar to Acts of Parliament,—and until all specifications are printed also and sold at similar prices,—such a confused and difficult state of things must, it is feared, continue to remain, to the utter shame and disgrace of this inventive and mechanical nation.

Let us take, for example, the case of a party wishing to erect an electric telegraph of his own or of his friend's invention. How is it to be ascertained whether or not the plan is any infringement of at least one, if not of some dozen existing patents? Every patent relating to electric telegraphs, to the production of electricity, and magnetism, &c., must be carefully sought out; every word of every specification carefully read and compared with the existing state of public knowledge at the date of the patent, and read too with a full knowledge of the contents of probably many previous patents referred to therein.

Some specifications, instead of quoting from, refer to parts of other specifications. The specifications are written on parchment, and sometimes in characters scarcely legible by persons of the present day. The plans accompanying the specifications are often stitched yards and yards from the parts of the specification where they are referred to.

Before the specifications are copied on to long narrow rolls of parchment, many have occupied (exclusive of plans) 16 to 18 skins of closely-written parchment. Not a word may be copied at the time of the reading of the specification in the Office—nor will the officials copy for payment *a portion only* of the specification. No pencil or paper is allowed. The reader must carry in his head the meaning, the bearing, and the contents of the whole specification, and of every single sentence in some 16 or 18 skins of parchment, and that too of perhaps some 20 to 30 specifications, before he can feel at all confident in his own mind that the erection of the telegraph he desires will not bring down upon him some half-dozen actions in law from various quarters.

Such is the glorious uncertainty of the matter, and such are the facilities at present afforded to inventors and other ingenious and scientific men of the day, to enable them to steer clear of the breaking of the law, and to give to the public new and improved plans of telegraphs.

The author having gone through all the nuisance of this searching and reading of the specifications of every patent for electric telegraphs, and having endured the horrible torture of learning claims by heart, and of filling his head to the full in the Enrolment Office, and emptying it on paper immediately he has emerged from the door of those dreaded precincts, can well bear witness to the abominable and disgraceful state of things as regards the present state of the patent laws in this kingdom.

But enough,—it is certain that the reader will not wish to have an extended account of telegraphs, the description of which has to be acquired by such means, nor will he place much reliance on the more minute details obtained from specifications in the manner above given.

It is proposed therefore, on all the above grounds, to give only an *abstract* of a few of the *leading features* of some of the telegraphs patented from the year 1837 to the present time, selecting the patents of those parties whose inventions are now in daily practical operation in the kingdom.

The author would observe, that he has not hesitated to put into print long ago the specifications of the various patents taken out by himself.

It is now proposed to deviate slightly from the plan heretofore followed, viz. of giving the inventions of different parties merely *in the order of the dates* of such inventions. It is believed that when a person takes out a patent one year, and then improves upon his invention and so patents this second improvement in the next year, and so on with a third, fourth, and fifth set of improvements, for which separate patents have been respectively taken out, that it will be much better to *continue* the description of *those patents* in the order of their date, than to leave off and to commence the description of the inventions of other parties—inventions relating perhaps to a wholly distinct class of telegraphs.

7
TELEGRAPHS FROM 1837 TO THE PRESENT DAY

Cooke and Wheatstone's Telegraphs

On the 12th of June, 1837, Messrs. Cooke and Wheatstone took out letters patent in England for "improvements in giving signals and sounding alarums in distant places by means of electric currents transmitted through metallic circuits."

Many persons have long had an idea that Messrs. Cooke and Wheatstone were the first inventors of *the electric telegraph*. It is clear, however, from what has been said before, and from the very title of this patent, which is for *improvements*,—and those improvements relating to certain particular parts only of the electric telegraph,—that such a notion is wholly erroneous.

The peculiar features of this telegraph were, as they are expressed in the title, *improvements* on the well-known modes of making the signals and of sounding alarums.

Five magnetic needles and coils were used, and either five or six wires employed, accordingly as it was desired, by means of the needles, to produce twenty or thirty primary signals. The needles were arranged in a horizontal row and on a vertical dial, and steps were placed to cause each needle to remain inclined at a particular angle when acted on by electric currents.

Letters were engraved on the dial at the points where the lines of convergence of two needles met; by causing two of the five needles to converge, a letter could be denoted. Five wires and five needles gave twenty of such signals; if a sixth wire were employed, but without a needle, then only one of the five needles could, if desired, be moved; and thus, by the single motion of each of the five needles to the right or left, ten other signals could be given. The improvement relating to the alarum was the employment of the attractive force developed in soft iron (when electric currents were caused to pass round it in coils of wire) for the purpose of striking a bell or releasing wound-up mechanism.

It is clear that this telegraph, as a whole, was a great improvement on many others at that day, though still very far from perfect.

The peculiar arrangement of the dial at once reduced the number of wires which would have been required under Ampére's plan from twenty to five, although it must not be forgotten that at this time many one-wired telegraphs were well known.

A peculiar kind of keyboard was employed, and other mechanical improvements effected. It must be observed, however, that this telegraph contains little or nothing new beyond *the peculiar combination* of well-known parts. The use of the needle and coil was old; the employment for telegraphic purposes of galvanic electricity was old; the burying of insulated wires in tubes was old; the attractive force of soft iron to develop electro-magnetic properties was old. But the peculiar mode of *"giving signals and sounding alarums"* (the words as given in the title) was new, and was an improvement on the then known plans of this class of telegraphs, and a *great improvement* too.

The author has been anxious to put this matter in what he considers its true light, as much misapprehension has arisen as to what the real inventions in this patent were,—a misapprehension which he conceives has arisen from the great difficulty which all persons (excepting only those who have spent many years in this particular branch of science) experience in obtaining a correct knowledge as to what was and what was not the common stock of knowledge possessed by many parties at that date conversant with the science, and especially as to what had been already done in electric telegraphs.

To Professor Wheatstone himself credit must be given, not only of knowing what had been done in this country, but what had been proposed and done in almost every civilized country on the Continent, both as regards electricity and electric telegraphs; and hence the reason is obvious why the first patent taken out for an electric telegraph in this kingdom had in its title, for its security's sake, the word "IMPROVEMENTS."

The author conceives that great injury has been done to Professor Wheatstone and his partner Mr. Cooke by parties claiming for *them* the first invention of *the* electric telegraph; whereas if those friends had but read the first published words of the inventors themselves, they would have found that all that they themselves had said was, that what they had invented were only certain *improvements*. Much undeserved bitterness and acrimony of feeling have thus been raised unjustly against the first *patentees* of improvements in electric telegraphs.

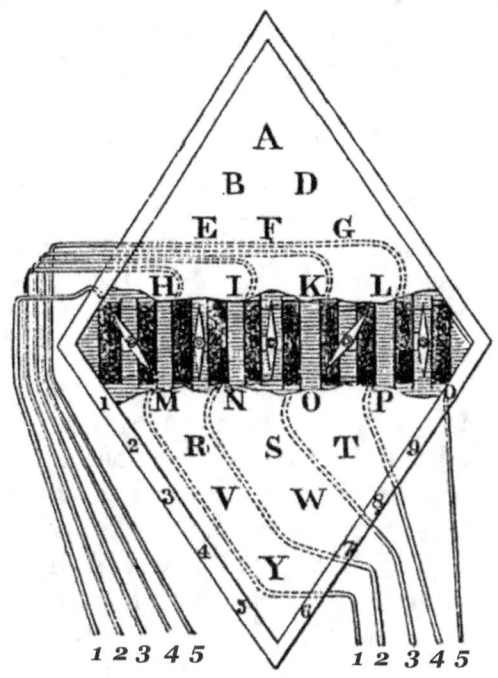

Fig. 16
Cooke & Wheatstone five-needle telegraph

COOKE AND WHEATSTONE'S FIVE-NEEDLE TELEGRAPH
The following description will explain this form of telegraph, patented in 1837. Cooke and Wheatstone's arrangement required the service of five galvanometers. Figure 16 is a representation of the dial.

In the interior there are five galvanometers, numbered 1, 1;

2, 2; 3, 3; 4, 4; and 5, 5. The coils of the multipliers are secured with their needles to the case, having each exterior needle projecting beyond the dial, so as to be exposed to view. Of the wires from the coils, five are represented as passing out of the side of the case, on the left hand, and are numbered 1, 2, 3, 4 and 5. The other five wires pass out on the right hand, and are numbered in the same manner. The wires of the same number as the galvanometer are those which belong to it, and are continuous. Thus the wire 1, on the left hand, proceeds to the first coil of galvanometer 1, then to the second coil, and then coming off, passes out of the case, and is numbered 1, on the right hand; and so on with the other wires. The dial has marked upon it, at proper distances and angles, twenty of the letters of the alphabet, viz., A, B, D, E, F, G, H, I, K, L, M, N, O, P, R, S, T, V, W, Y. On the margin of the lower half of the dial are marked the numerals, 1, 2, 3, 4, 5, 6, 7, 8, 9, and 0. The letters C, J, Q, U, X, Z, are not represented on the dial, unless some six of those already there are made to sustain two characters each, of which the Specification is silent.

Each needle has two motions,—one to the right, and the other to the left. For the designation of any of the *letters*, the deflection of two needles is required, but for the *numerals*, one needle only. If one needle only were required to be moved, *six* wires were necessary. The letter intended to be noted by the observer is designated, in the operation of the telegraph, by the *joint deflection* of two needles, pointing by their convergence to the letter. For example, the needles 1 and 4 cut each other, by the lines of their joint deflection, at the letter V, on the dial, which is the letter intended to be observed at the receiving station. In the same manner any other letter upon the dial may be selected for observation.

Suppose the first needle to be vertical, as the needles 2, 3, and 5, then needle 4 being only deflected, points to the numeral 4 as the number intended to be signified.

SUBSEQUENT PATENTS BY COOKE AND WHEATSTONE
The next inventions of Messrs. Cooke and Wheatstone were patented in the name of Mr. Cooke only. The patent was sealed on the 18th day of April, 1838. The principal improvement in this patent

consisted in a *peculiar means* of enabling two intermediate stations to communicate with each other and with either terminus.

Under the first patent a message could be sent from either terminus, and it could also be read off at an intermediate station; but the intermediate stations could not, with the peculiar arrangement of the keys described in the first specification, *send* a communication to each other or to either of the termini.

The plan set forth would be extremely difficult to describe without a model; even the Specification, accompanied with its drawings, is quite unintelligible to the general reader. This plan, however, is not in use now, and has long been superseded by other later and better inventions.

Another plan of sounding an alarum is described in this patent. Wound-up mechanism was to be liberated and a bell sounded *by the angular motion of a magnetic needle,* the motion being produced in a manner similar to the motion of the needles of the telegraph instrument in the previous patent.

The peculiar form of telegraphic instruments, as described in these two patents, for sending intelligence, was tried both on the Great Western and the London and Birmingham Railways, but was soon abandoned, and has never since been used either in this or any other kingdom. A different form of telegraph, viz. an instrument having only two needles, is now in use, and in some cases only one needle is employed.

Figures 17 and 18 on the following page represents one of these double-needle instruments.

The following is an account of the number of words per minute sent by the double-needle telegraph in 11 despatches for the 'Times' newspaper, in the year 1849. The average per message is at the rate of nearly 17 words per minute:

364 words, at the rate of 13½ words per minute.
166   "   "   8¼"   "
383   "   "   14⅓   "   "
447   "   "   17¼   "   "
101   "   "   20⅕   "   "
288   "   "   17   "   "

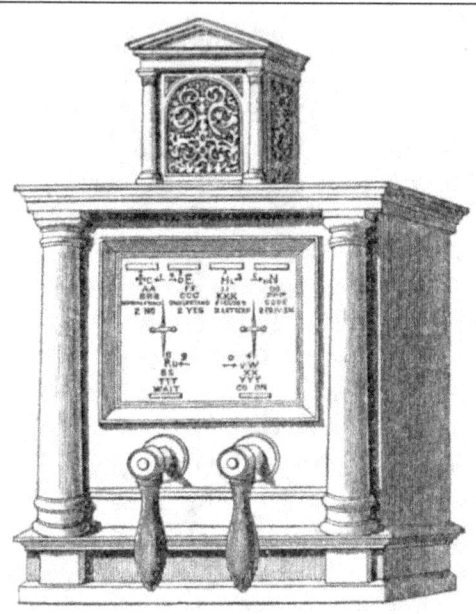

Fig. 17
Cooke and Wheatstone double needle instrument (exterior view)

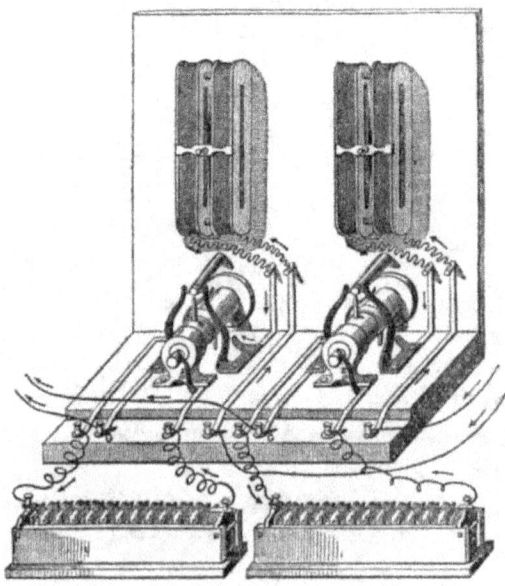

Fig. 18
Cooke and Wheatstone double needle instrument (interior arrangement)

274	"	"	15¼	"	"
106	"	"	15½	"	"
102	"	"	12¾	"	"
334	"	"	17½	"	"
73	"	"	18¼	"	"

The next patent taken out by Messrs. Cooke and Wheatstone was dated the 21st of January, 1840.

The telegraph included under this patent is a step-by-step letter-showing telegraph. A disc, having all the letters upon it, was fixed on the axle of a piece of clockwork mechanism,—an electromagnet and armature of iron were attached, so that as the armature was attracted to the electromagnet, the disc was allowed to progress one step forward in its revolution, and thus to expose to view, in a small orifice cut in the dial for the purpose, the letters of the alphabet, one by one at a time. A pallet and escapement wheel, similar to those employed in a clock, were used, so that every current of electricity transmitted allowed the disc to progress one step or tooth, and to expose to view successively each letter of the alphabet. When a particular letter was to be recorded, the disc was allowed to remain for a short period of time, with the desired letter, exposed to view.

Electricity, from a peculiar arrangement of the magneto machine, was also employed.

This telegraph is known as the Revolving Disc Telegraph. The great difficulty experienced with this telegraph was the impossibility of making the discs at two or more stations to move exactly together, *i. e.* for neither disc to lag behind.

If this lagging behind did occur, then when B was visible at one station, A would be visible at another, and ever afterwards all the letters would be wrong.

The Specification also describes another modification of an alarum. This telegraph has never been brought into any extended practical use in this kingdom, and is now entirely abandoned for other plans. Figure 19 will sufficiently explain the arrangement.

The next telegraph by Messrs. Cooke and Wheatstone was

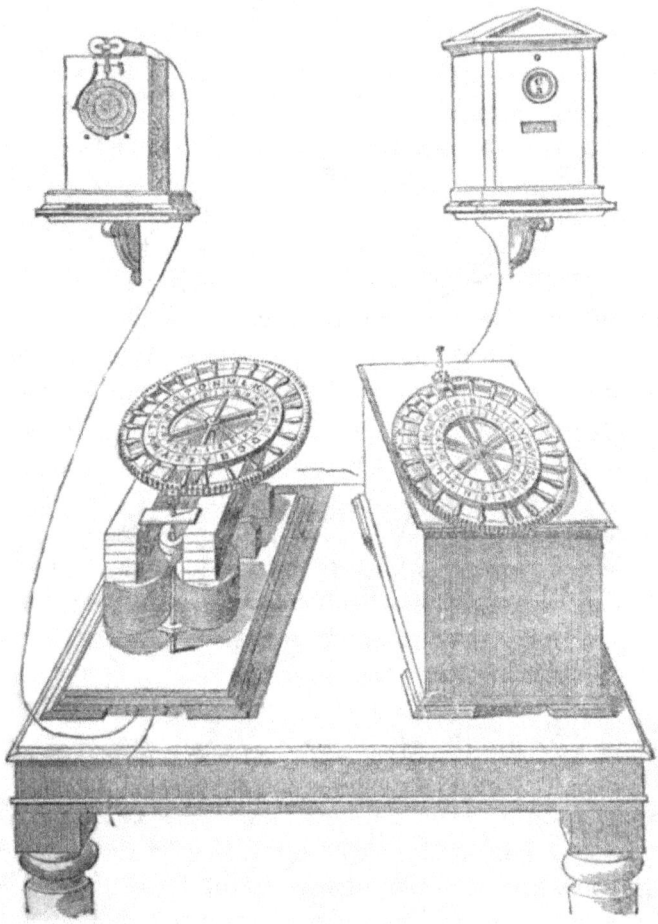

Fig. 19
Cooke and Wheatstone Electro-Magnetic Telegraph
(also known as the Revolving Disc Telegraph)
as tried on the Great Western Railway

patented on the 7th of July, 1841, in the name of Wheatstone only.

The Specification principally refers to electric engines and particular means of producing and developing the power; but it has also claims for parts having reference to electric telegraphs.

The parts referring to telegraphs contain descriptions of modes of making marks on paper by means of transfer paper, and modes of causing two or more electromagnets to act in *succession* by means of electricity sent over only one line-wire.

The next patent taken out by Messrs. Cooke and Wheatstone was for particular modes of suspending wires in the air. This patent bears date the 8th of September, 1842, and was taken out in the name of Mr. Cooke only.

The modes described are various, but the principal features were the causing of zones of dry wood to exist between wire and wire by means of artificial boxes or circular sheds like umbrellas,— the tightening of wires by certain well-known mechanical means,— the use of compound twisted wire—a kind of portable telegraph instrument to be attached to the wires,—as also the use of wires suspended under the particular modes as described and patented, if used for the purposes of sending currents of electricity to work electric clocks, or particular kinds of apparatus connected with certain descriptions of electric telegraphs.

The plan of causing *zones* of dry wood to intervene between wire and wire was tried and has been abandoned. It was succeeded by the following method, which has been very extensively employed in England until within the last few months.

The Figures 20, 21, 22 and 23 will explain this plan:

a a are arms of wood attached to a post or standard by means of a bolt passing through the porcelain tubes *y y*. *e e* are tubular insulators of porcelain, affixed to the arms by clips of iron. The wires pass through the tubes *e e*, and are thereby insulated. About every tenth post is made stronger than the intermediate ones, and strong cast-iron ratchet-wheels, with barrels, *r r*, are affixed to it for drawing up the wires. When the wire has been threaded through the insulators *e e* on the intervening poles, its end is attached to these winders, and on turning the ratchet

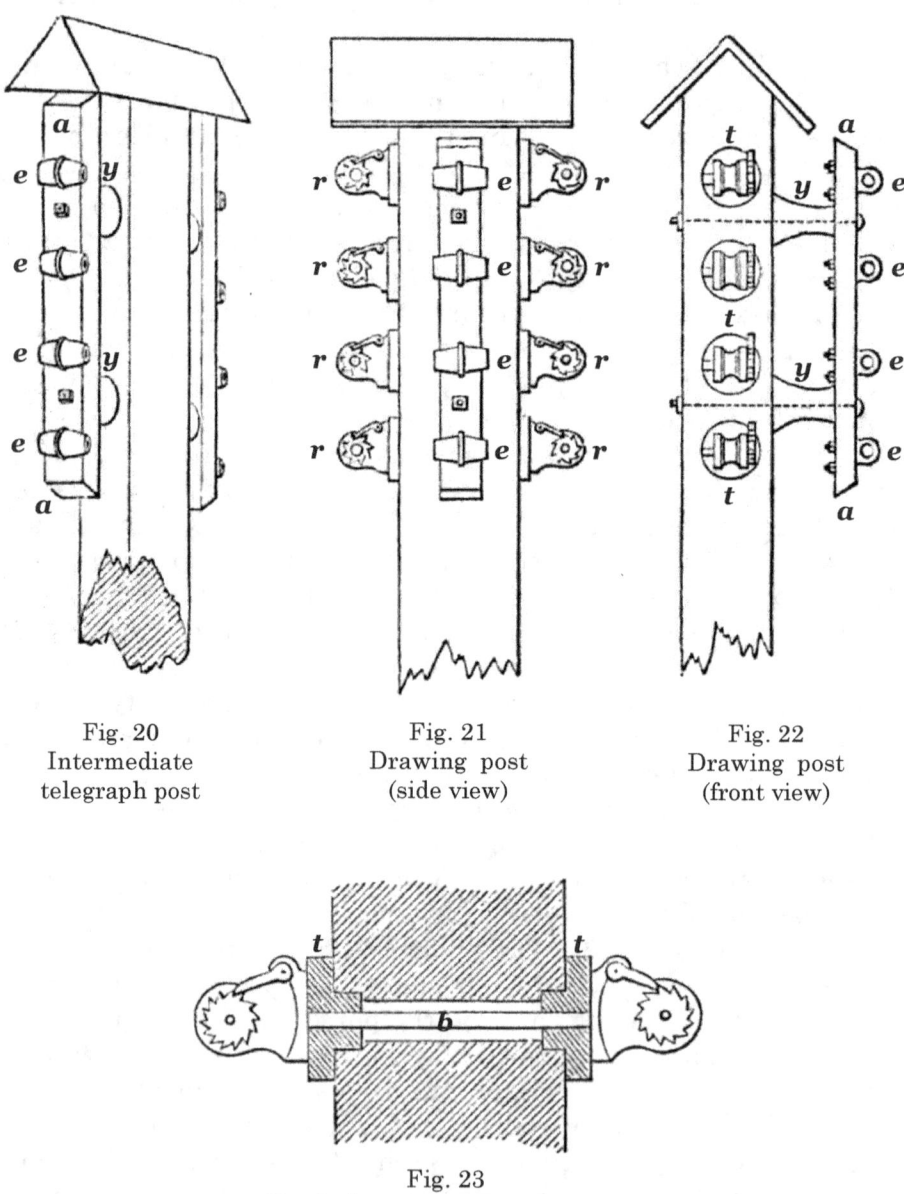

Fig. 20
Intermediate
telegraph post

Fig. 21
Drawing post
(side view)

Fig. 22
Drawing post
(front view)

Fig. 23
Vertical section of Figs. 21 & 22

Telegraphs from 1837 to the Present Day

LENGTH	AT BASE		AT TOP	
	Drawing Posts	Intermediate Posts	Drawing Posts	Intermediate Posts
18 ft.	9 in. × 8 in.	6 in. × 6 in.	7 in. × 6½ in.	5½ in. × 4½ in.
22 ft.	10 in. × 8 in.	7 in. × 6 in.	ditto	ditto
28 ft.	11 in. × 8 in.	8 in. × 6 in.	ditto	ditto

Principal dimensions of the first telegraph poles erected in England

wheels round by means of a strong handle, the wire may be wound round these barrels and thus drawn up to any degree of tension desired. The ratchet wheels and barrels on each side of the post are connected to each other by the bolt *b*, and are insulated from the post by means of the porcelain tubes *t t*. Figures 21 and 22 show the view of a drawing post with these winders attached, and Fig. 20 the view of an intermediate post.

Figure 23 is a vertical section of Figures 21 and 22. The posts are shown fitted up for two vertical rows of wires. The wires now used are of iron, which is galvanized, to protect it from the action of the damp atmosphere. They are of about one-sixth of an inch in diameter, corresponding to No. 8 of the wire gauge. The wire being obtained in as great lengths as possible in the first place, successive pieces are welded together until a length of about 440 yards has been formed, the weight of which is about 120 lbs.

The Table above shows the principal dimensions of the posts and poles used in the first telegraphs of the kind erected in England.

This mode of suspending wires is now, however, being abandoned for a more simple and inexpensive method. In the first place, the poles are of larch or common fir,—the winding apparatus is dispensed with,—and a new form of insulator adopted.

The first plan adopted by Messrs. Cooke and Wheatstone in the extension of the conducting wires between distant points, was to cover each wire with cotton or silk, and then with pitch, caoutchouc, resin, or other non-conducting material, and to enclose them, when thus insulated, in tubes or pipes of wood, iron, or earthenware. The telegraph on the Great Western line was originally laid down on this method. This was soon abandoned for the introduction of wires on poles insulated by *zones* of dry wood. This second plan in its turn has also been discarded, and the wires and poles have been pulled up.

Excepting in localities where the suspension of wires is difficult, as in streets and towns, or on public roads, the earlier method of placing the wires in tubes of iron or wood, as practised by Messrs. Cooke and Wheatstone and others, has given place to the above plan.

The last patent taken out by these parties was dated the 6th of May, 1845.

The patent is very voluminous and contains several improvements in the detailed or dissected portions of the telegraphs then in use. To give an idea of the length of this Specification, it is necessary only to state, that the copy which the author has occupies no less than 90 sides of foolscap paper, written on very closely, and with not less than 30 lines to each page, *i.e.* 2,700 lines of closely-written foolscap.

The claims amount to fourteen in number.

The improvements relate to particular modes of moving magnetic needles,—modes of arranging stops to needles,—modes of arranging pointers,—modes of producing audible sounds for particular purposes,—a particular kind of code,—other modes of moving pointers,—a mode of attaching a portable telegraph to the line-wires,—improvements in galvanometers—a mode of setting free an alarum by a falling weight,—covering iron wire with leaden tubes when the tubes are to be suspended in the air,—alterations in magneto-machines, and lastly, a particular kind of key apparatus.

An Act of Parliament was obtained in 1845, for incorporating a Company under the title of "The Electric Telegraph Company," for the purpose of working these patents.

DAVY'S TELEGRAPH

The patent taken out next after the first two patents of Messrs. Cooke and Wheatstone was by Mr. Edward Davy, and was sealed

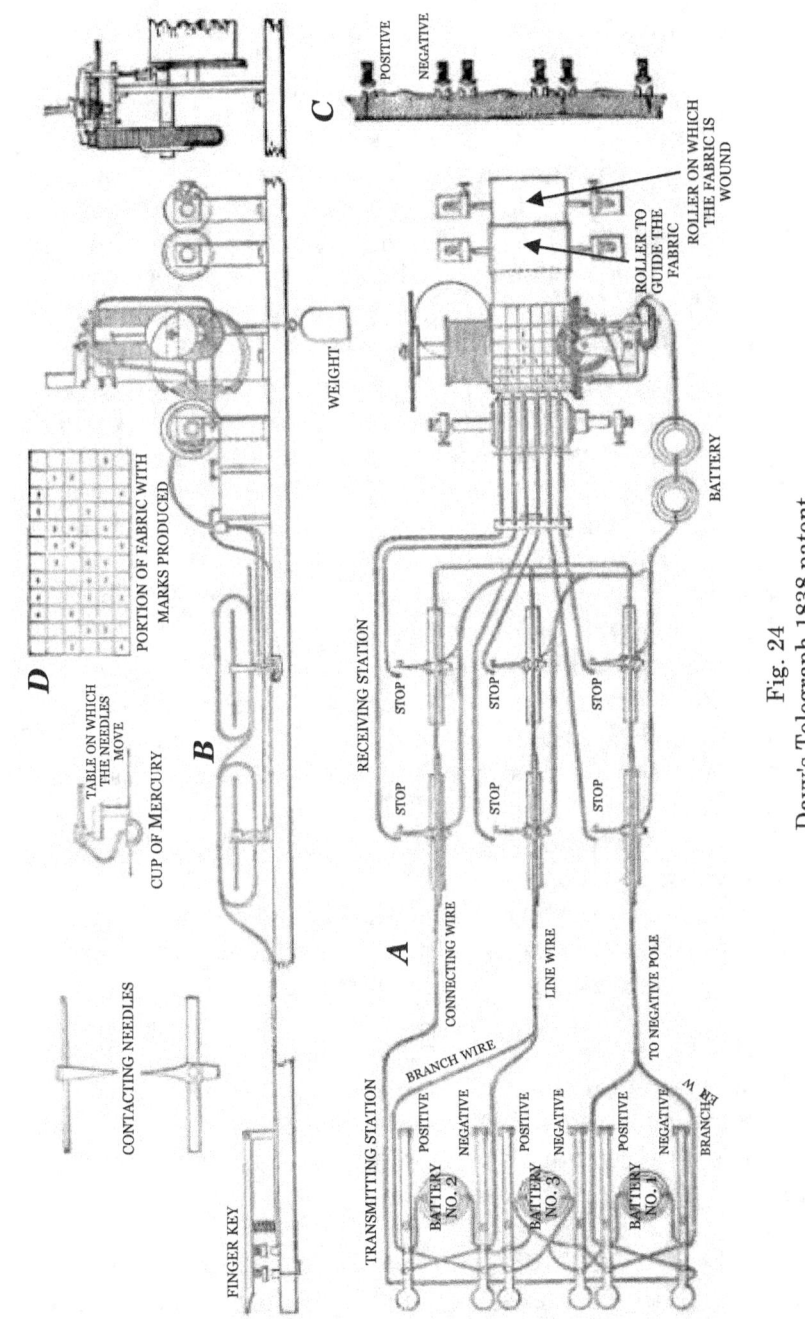

Fig. 24
Davy's Telegraph 1838 patent

on the 4th of July, 1838.

This patent was for a chemically marking telegraph. Three wires were to be used, and points of metal wire were to be caused to press, by means of the motion of magnetic needles, upon chemically prepared fabric at the distant or receiving station.

The fabric to be employed was calico or paper, and it was to be moistened with a solution of hydriodate of potass and muriate of lime.

The motion of a needle to the right caused a mark to be made on *one* part of the fabric, and the motion of the same needle to the left caused a mark to be made on *another* part of the fabric; and the same for *each* needle attached to the respective wires. Thus the single or combined marks were made to express letters or other desired symbols.

The Specification contains many other improvements in the details of the telegraph. The form of telegraph, as given, has never been brought into practical use. The inventor left this country for Australia, and the patent has been bought by the old Electric Telegraph Company, but not used.

This patent would not have been noticed, if it had not in a great measure formed the *basis* of some of the chemically marking telegraphs as used in this or other countries; viz. those of Bain, Bakewell, and others.

A is a top view of the telegraph.
B is a side view.
C is a section showing the construction of the keys.
D is a drawing of the fabric as it would appear when marked by the colouring produced by the electric current.

BAIN'S TELEGRAPHS

We will now pass on to the patents of Mr. Alexander Bain. Mr. Bain, in 1845, opposed the Bill of the old Electric Telegraph Company, when before Parliament. The result was a compromise between the parties, and the purchase by that Company of the patents of Mr. Bain.

The first patent of Mr. Bain was sealed December 21, 1841. This patent relates to a telegraph applicable to locomotive engines.

With regard to that part which has more immediate reference to the electric telegraph, Mr. Bain proposed to have the coil moveable, and the magnet stationary; Cooke and Wheatstone's plan being the reverse of this,—viz., the coil being stationary, and the magnet moveable.

Mr. Bain proposed to apply this mode of obtaining motion both to ordinary telegraphs as well as to a new form of printing telegraph. This printing telegraph was an extremely ingenious one at the time. A modification of it was afterwards at work for some time over a few miles on the South Western Railway.

In this patent also was included a mode of insulating wires by means of bitumen.

A second patent was taken out by Mr. Bain in 1843. This patent contains improvements on the foregoing plans, besides several other new arrangements, and also a plan of lowering the plates of a battery by means of clockwork mechanism and an electromagnet, so as to keep the power employed always of the same strength.

The patent has also inventions with respect to electric clocks, and describes as well a mode of producing copies of type by means of electro-chemical decomposition.

The telegraph known as Bain's I and V telegraph (so called from the particular figures which were employed in forming words and sentences) is fully described in the Specification.

Mr. Bain also, in the same Specification, describes his mode of burying in the earth a mass of copper at one terminal station, and a mass of zinc at the other, and joining, when desired, these metals by the line-wire. A current of electricity could of course flow through the wire, and this electricity was to produce the necessary signals.

To those versed in the science of electricity, it will be evident that this arrangement was but the using of one large cell, in which the mass of copper and of zinc formed the plates, the earth the jar or cell, and the moisture in the earth the exciting liquid.

The practical objections to this arrangement consist in the want of intensity in the electricity generated, and in the motion of a magnetic body being obtained only in one direction.

A third patent was taken out by Mr. Bain in 1845. This patent was sealed on the 25th September.

The first part refers to suspending wires in a kind of fence railing, and also a peculiar mode of suspending wires on posts.

Another part refers to the handle apparatus of a telegraph for transmitting currents of electricity. There is nothing new in the principle herein employed, but the mechanism differed from that in use at the time.

There are also modes of sounding alarums, and also improvements on codes to be used with the I and V telegraph.

Improvements are also set forth in step-by-step movement telegraphs, and in printing or dotting telegraphs. Several improvements in electric clocks are also described.

A fourth patent was taken out by Mr. Bain in 1846. This patent was sealed on the 12th of December. The first part refers to the mode described as a one-wire chemically marking telegraph. A circular wheel, with moveable projecting pins, was employed. When the pins were pulled out as the wheel revolved, they came into contact with other spring pins, and thus caused currents of electricity to be transmitted from a battery, producing thereby corresponding chemical marks on chemically prepared paper at the distant station.

Another plan consisted in cutting out slits of different lengths in a long strip of paper at the transmitting station, and allowing this perforated strip to pass uniformly over a metal cylinder with a pin or spring pressing on the top of the paper. When ever, therefore, a hole in the paper passed under the pin, the pin came into metallic contact with the cylinder underneath, and allowed a current of electricity to pass through the line-wire. All the holes in the paper, and their length, were therefore proportionally represented at the distant station by chemical marks of corresponding lengths on the prepared paper at that station. This form of telegraph is the quickest at present invented. It does not, however, seem suited to ordinary communications, but only to the transmission of *very* long documents on extraordinary occasions.

If one person only is employed to punch holes in the paper, it is evident that, instead of making a hole in the paper, a current of electricity might as readily be sent, and a chemical mark made at the distant station, and thus the message might actually be sent in the same time as that required for cutting the paper. But this remark applies only to the case where there is but one attendant for a wire. If a *number of men* be employed at each station, then, by dividing the message into parts, and each man punching out his part, the whole paper can be perforated in less time than *one* man could send the message. On uniting this perforated paper, and applying it to a machine, and on turning the cylinder round, corresponding chemical marks may be made at a distant station with very great rapidity. The commercial question is therefore, where ordinary communications are alone required, one of large working expenses versus a rather larger outlay of capital in the first instance. The plan proposed, however, is most ingenious, and the instrument will form a good adjunct to the other instruments at very important stations.

This same patent also includes a form of telegraph post. This is composed of four thin slabs of timber fastened together in the manner of a box, the interior being hollow. The author is not aware that this plan of post has ever been adopted.

Owing to the fact that no publication of the Specifications of these patents has yet been made, the author is unable to give drawings of the same, or to refer more fully thereto. These telegraphs show very great ingenuity in their various parts, as also in the mechanical details employed.

H. & E. HIGHTON'S TELEGRAPHS

The next patents in order are those of the Messrs. H. and E. Highton,—viz. those of the Rev. H. Highton, M.A., and of his brother, Mr. Edward Highton, C.E., (the author of the present treatise.)

The first patent was taken out in 1844 by the Rev. H. Highton. In this telegraph electricity of high tension was employed, viz. that produced either from the ordinary electric machine, or from the hydro-electric machine: one wire only was used. A piece of paper, which was moved uniformly by clock-work mechanism, was conducted at the receiving station between two points of metal in connection with the line-wire, the points being placed one *above* the other, and on *opposite* sides of the paper. On sending currents of electricity, the paper was pierced by the electricity, every shock making a little hole through it. If the electricity transmitted were positive, a hole was pierced at *one* of those points, and if negative, a hole was made at the *other* point. By the combination of these perforations letters and symbols were denoted.

By an arrangement of these dots or holes, under the ordinary mathematical law, from 30 successive currents of electricity, occupying, say, 15 seconds of time, no less than 1,073,741,824 different signals could be made.

Ten miles of wire were erected on the London and North Western Railway, for the purpose of testing the practicability of the plan, and of obtaining certain fundamental laws as to the transmission of electric currents. The signals were found to be given with great certainty, and the paper, moistened with dilute acid, was pierced even when a Leyden jar, filled only with water, and in size not greater than one's little finger, was employed.

The acceptance by the author of an appointment as General Superintendent to a railway company, and his being engaged in other engineering works, prevented the further carrying out of this plan of telegraph.

The plan was submitted to the Government, and an offer was made to connect Liverpool with London by means of this telegraph, and that at the sole risk of the patentee and the author, provided that the Government would obtain for them, for such purpose, liberty to use the lines of the London and Birmingham, Grand Junction, and Liverpool and Manchester Railways. The Government, however, found that at that time they possessed no compulsory power to grant such license, even for a telegraph for their own use; and hence in a Bill passing through Parliament at the time with reference to railways, clauses were added, giving this power to Government for telegraphs for their own purposes. This, it is believed, was done at the instigation of the late Sir Robert Peel, with whom the author was at the time in communication on the subject.

The paper, when marked, would thus appear is in Figure 25. This plan as shown, would correspond with the number

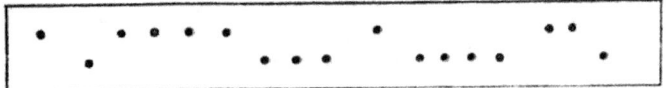

Fig. 25
Highton's system of marks caused by high-tension electricity

12,413,411, and would in sending occupy only some 5 or 6 seconds.

HIGHTON'S TELEGRAPHS

The next patent was taken out by the Rev. H. Highton, M. A., in 1846. The telegraph included in this patent is known as the Gold-leaf Telegraph.

A small strip of gold-leaf inserted in a glass tube was made to form part of the electric circuit of the line-wire. A permanent magnet was placed in close proximity thereto. When a current of electricity was passed along the line-wire, the strip of gold leaf was instantly moved to the right or left, according to the direction of the current.

The author lays no claim to this invention; he was at the time occupied in important engineering works, and did not even see this form of telegraph until it was bought by the old Electric Telegraph Company, in consideration of a small annuity for fourteen years, which the Company agreed to pay to the inventor for the exclusive use of the invention.

The following Specification will fully explain this most simple and perfect telegraph, illustrated in Figure 26:

> Extract from the Specification of the Patent granted to Henry Highton, of Rugby, in the county of Warwick, Master of Arts, for improvements in Electric Telegraphs. Sealed February 3, 1846.
>
> To all to whom these presents shall come, &c. &c.—In the electric telegraphs now commonly used on English railways, signals are given by the motions of magnetic needles, which are caused to move to either side by the action of electric currents passed in either direction through coils of wire surrounding magnetic needles. And I have discovered that signals can be exhibited in electric telegraphs by motions produced by electric currents in strips of metallic leaf, suitably placed, in a very cheap form of signal apparatus, resembling a gold-leaf galva-

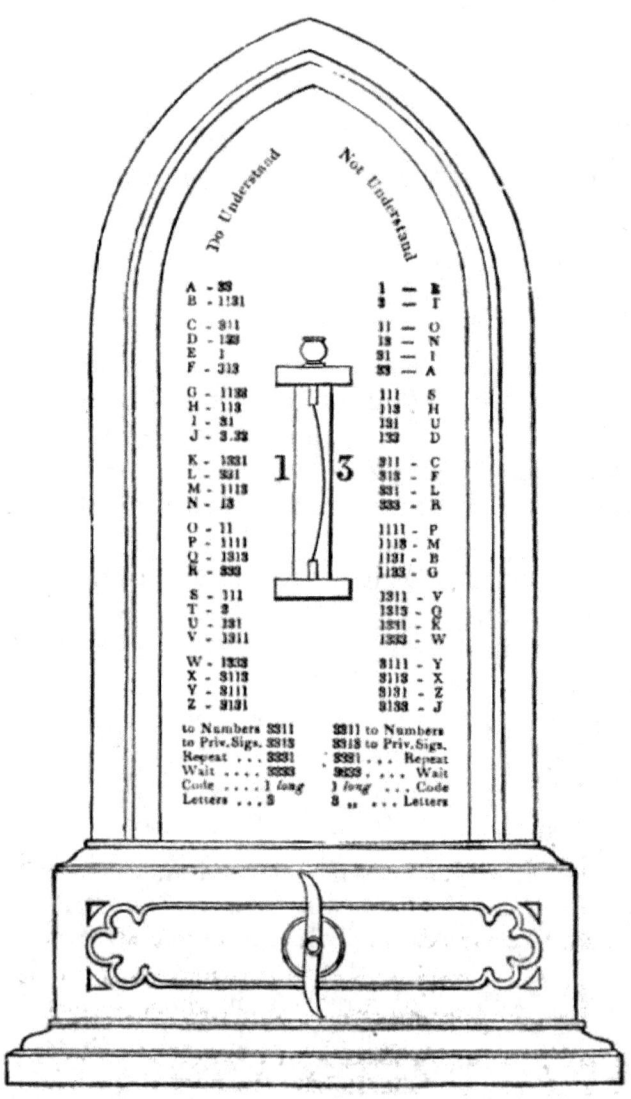

Fig. 26
Highton Gold-Leaf Telegraph for one line-wire, with code table shown on dial

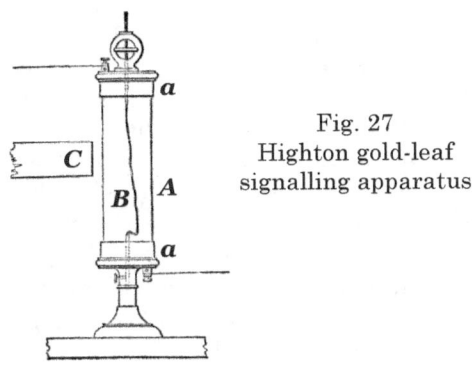

Fig. 27
Highton gold-leaf
signalling apparatus

nometer.

The drawing hereunto annexed (Figure 27) represents a signal apparatus, consisting of a glass tube, *A*, fitted in brass caps, *a a*, at top and bottom, and having a strip of metallic leaf, *B*, (gold leaf being the kind of metallic leaf which I usually employ,) passing through its centre, loosely hung, in metallic contact with the said caps; the upper extremity of the metallic leaf being fixed at right angles to its lower end, so that the metallic leaf, from whatever direction seen, will present at some part its flat surface to the eye. The caps, *a a*, (which are removable, in order that the metallic leaf may be replaced, if broken,) are placed in a circuit suitable for electro-telegraphic communication.

Near to the metallic leaf (as on the outside of the glass) is placed either of the poles of a magnet, *C*. And the effect of this arrangement is, that when a current of voltaic electricity is caused to pass through the circuit, and, therefore, also through the metallic leaf, *B*, included in it, the metallic leaf is deflected to one side or the other, according to the direction of the current. And the distinct motions so obtained may be repeated and combined, and used for the purpose of designating letters or figures, or other conventional signals.

One of the above-mentioned signal apparatuses is placed at each terminus of telegraphic communication, and others may be placed at intermediate points.

Each terminus, and also each intermediate station, is provided with a voltaic battery and with one of the keyboards in use in single magnetic-needle electric telegraphs. The person in charge of the telegraph at either terminus, or at any intermediate station, produces the requisite connections for causing an electric current to pass in either direction through the circuit, and, therefore, through the metallic leaf of the signal apparatus of each terminal or intermediate station, and thus cause the metallic leaf of all the signal apparatuses to move simultaneously to either side, so as to give the required signal or signals.

The keyboard of each terminal or intermediate station has a handle, by moving which the person in charge of the telegraph at any station can cause an electric current to pass through a circuit in connection with a system of alarums at the terminal and intermediate stations, similar to those in use in magnetic-needle electric telegraphs.

The next patent was taken out in January, 1848, by Messrs. H. and E. Highton. To the inventions contained in this patent, the author devoted a great amount of labour, money, and time.

The author was acting as Telegraphic Engineer to the London and North Western Railway Company, and was pressed by that Company to invent a set of electric telegraphs free from the objections and defects inherent to most telegraphs then in use, and free also from any of the then existing patents.

Every telegraph proposed or executed, either at home or abroad, was minutely investigated, and their defects studied with the greatest care. Neither time nor money was spared to accomplish the objects desired.

The result was a series of inventions of great variety and extent. For these inventions, the patentees received from the hands of His Royal Highness Prince Albert, as President of the Society of Arts, the greatest honour the Society has the power to bestow, viz. their Large Gold Medal.

Several of the plans were immediately adopted on the London and North Western Railway, in preference to those of the old Electric Telegraph Company, who then possessed a great number of patents. The telegraphs gave the greatest satisfaction, and have been in constant daily use ever since.

But to enumerate the principal features only of the inventions in this patent:—

- The horse-shoe magnet was suited to coils, and found to be

Telegraphs from 1837 to the Present Day

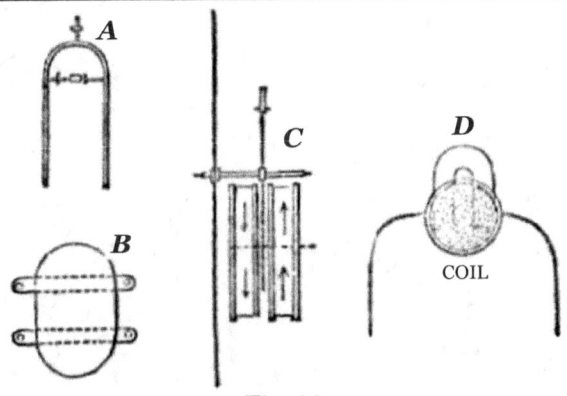

Fig. 28
Highton patented telegraph arrangements

much superior to the old straight magnetic needle and coil of Messrs. Cooke and Wheatstone.

- In step-by-step motion telegraphs a means was provided for causing the pointer or disc at once to progress by one bound to zero on the starting-point.
- The maximum work capable of being produced by any number of lines was taken advantage of, and thus three wires were made to produce 26 *primary signals*, and so to show instantly any desired letter of the alphabet.

Under Ampère's plan, 26 wires must have been used, and under Cooke and Wheatstone's patent 6 wires.

- Suitable keys were devised for sending currents of electricity over three wires in the 26 orders of variation.
- Direct-action printing telegraphs were devised, so that a single touch of one out of 26 keys caused instantly any desired one out of 26 letters or symbols to be printed.
- The insulation of wires was improved, and many other improvements relating to electric telegraphs effected.

Figure 28 above shows some of the telegraph arrangements constructed under this patent.

A, *B*, and *C*, show one form of the arrangement and parts of the horse-shoe magnet and coil, as arranged under this patent. *D* is another form—in this the coil is circular.

The advantage of the horse-shoe magnet over the straight

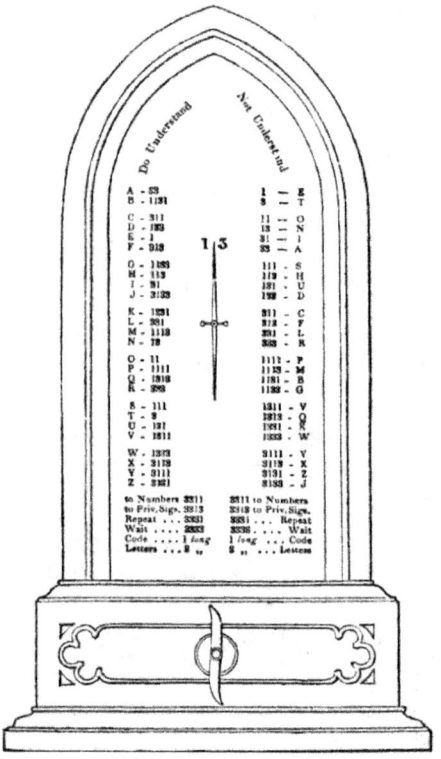 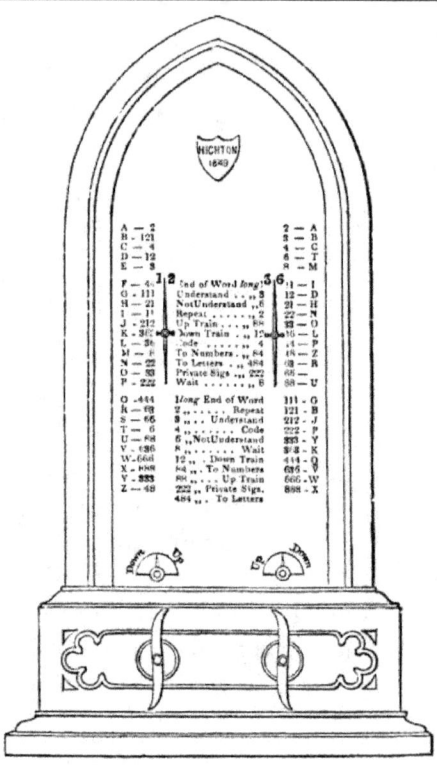

Fig. 29
Highton single-pointer telegraph for one line-wire, with code shown on the dial. The pointer is moved to the right or left by a horse-shoe magnet and coil.

Fig. 30
Highton double-pointer telegraph for two line-wires, with code table.

magnet or magnetic needle of Professor Wheatstone may be thus stated: When a coil surrounds a straight magnetic needle, as used by Messrs. Cooke and Wheatstone, *each* convolution of the wire has to pass *twice* over the central or *dead part* of the magnet; whereas, if the horse-shoe magnet be employed, there is *wire only* where there is magnetism in the magnet to be acted on.

This latter arrangement therefore enables all superfluous resistance in the circuit to be dispensed with; and hence the same amount of electric power is enabled to produce a far greater effect on the distant telegraphic instruments, or *less* power to produce an *equal* effect. Currents of electricity from secondary batteries

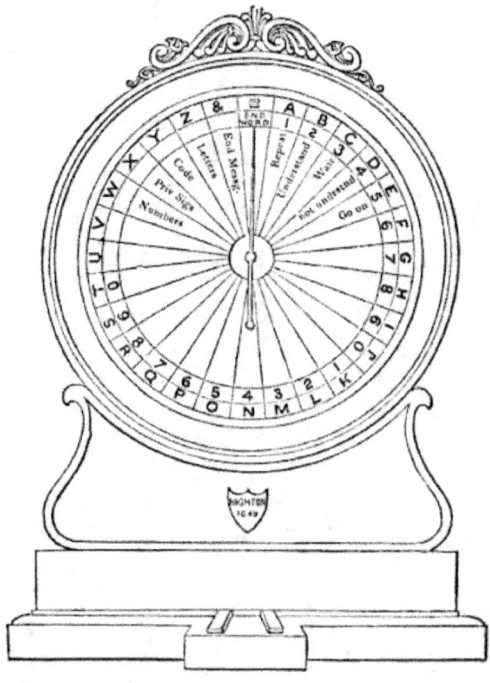

Fig. 31
Highton Revolving Pointer Telegraph with double-action escapement for either one or two line-wires, the pointer being able to progress from letter to letter, or to pass by one bound from any letter the whole distance up to zero.

were to be employed where great mechanical effects were desired at the distant station. An instrument was devised for this purpose, which was called a Peraenode.

Figures 29 and 30 (previous page), 31 above, and 32 to 36 on the following pages, illustrate some Highton Telegraphs and some printing mechanisms.

The next patent was taken out by the author on the 7th February, 1850. The patent contains a great many improvements in different classes of telegraphs. A few only of the principal features will be alluded to here. The first part refers to modes of arranging electric circuits.

Means of employing electricity of different degrees of tension

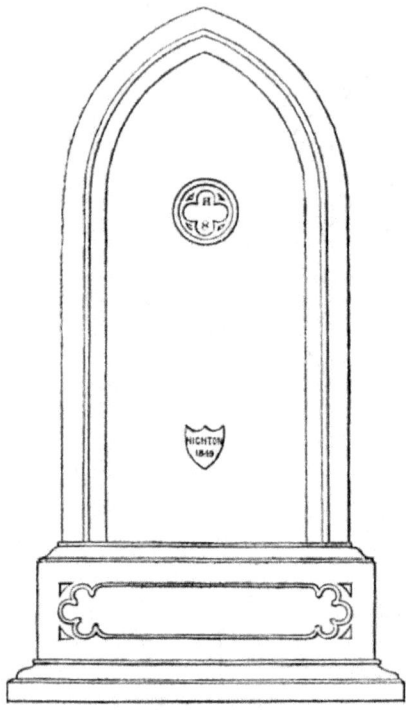
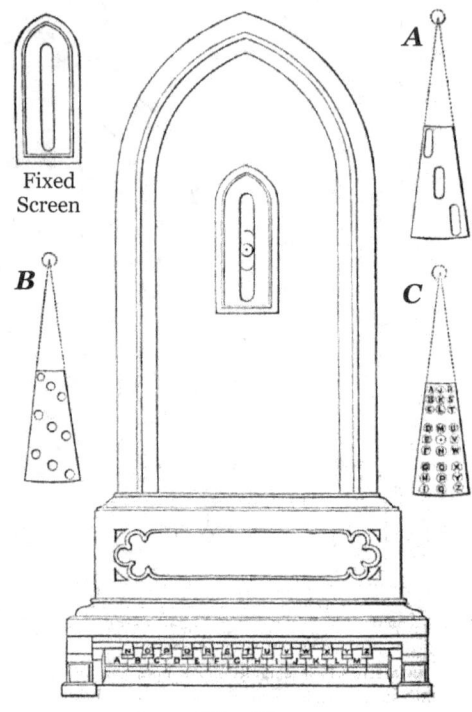

Fig. 32
Highton Revolving Disc Telegraph with new double-action escapement for either one or two line-wires.

Fig. 33
Highton Direct Letter-showing Telegraph for three line-wires. In this instrument the desired letters are brought instantly into view in the centre of the dial by means of three moveable screens, **A**, **B** and **C**.

and of different periods of duration are also shown, so that two kinds of electric apparatus may be connected to one line-wire, and one only worked, as desired. By this means one of the wires usually employed was rendered unnecessary. Other improvements relating to the dials are also made. A new mode of causing motion in soft iron, by temporarily magnetizing it by the contiguity of a powerful magnet, is described, which promises to be of great value in electric telegraphs, as by the employment of this apparatus any demagnetization of the magnets in thunder-storms is entirely obviated, and the coils of wire are made to give out more power.

Pendulous or *vibrating* bodies in step-by-step motion telegraphs

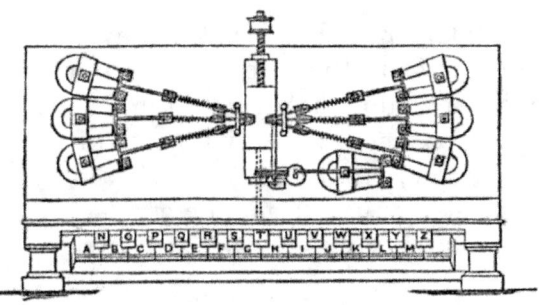

Fig. 34
Highton Printing Telegraph mechanism suited for either one, two or three line-wires, according to the rapidity of transmission desired. In this telegraph the letters are printed by one touch of a key when three wires are used.

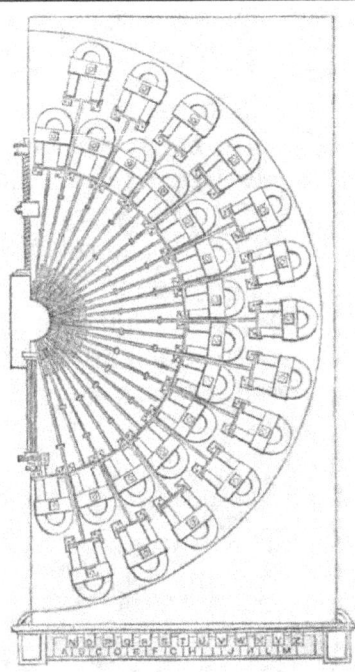

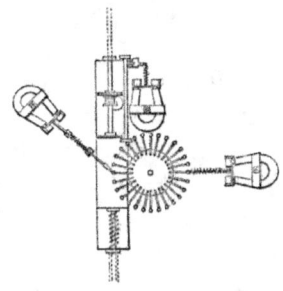

Fig. 35
Highton Telegraph mechanism for printing the letters of the alphabet, to be used with one line-wire.

Fig. 36
Highton Telegraph mechanism for printing letters of the alphabet, and suited to either one, two or three line-wires. In this telegraph the letters are printed by one touch of a key.

are introduced in order that a definite period of time may elapse between each successive current of electricity; and these same bodies are caused to make and break the circuit, so that no second current can be transmitted till all the instruments in a series have completed the work due to the prior current. In this way all overrunning or lagging behind of one instrument, as before described, is entirely obviated.

Another important improvement consists in the batteries. Batteries, as therein described, require not the slightest attention for months together, and many are now employed in doing the most severe work on the London and North Western Railway, which are never touched from periods of from three to twelve months at a time, and yet give out, whenever required, a constant and equable flow of the electric power. This is accomplished by the substitution of a solution of the sulphates of the earths instead of sulphuric acid. A further improvement consists in the formation of telegraphic posts, whereby those of the best foreign timber may be constructed at one half the former cost.

Many other improvements are described, which it would be tedious to enumerate, and which can only be properly understood by a reference to the Specification.

The next patent was taken out by the author in September, 1850. This patent relates to Submarine Telegraphs. The insulated wires are protected by being placed in the centre of a cable of wire-rope, or within strands of wire. Owing to the expense of patents in England as compared with their cost in France, and as it did not appear to the author that this arrangement of protecting the wire would be required in England, excepting between England and France, and England and Ireland, the patent was taken out for France, Ireland, and Scotland only, instead of for England as well.

For details of these several patents the reader is referred to the Specifications.

An Act of Parliament was obtained in 1850 for the incorporation of a Company under the title of "The British Electric Telegraph Company," for the express purpose of working and bringing into more general use the telegraphs of Messrs. Highton. The Printing Telegraphs of Messrs. Highton have not as yet been used for commercial purposes, as the wires of the British are only now in process of being laid down.

A brief extract from the Specification of the Telegraphic Wire-rope may not prove uninteresting.

> Extract from the Specification of the Patent granted to Edward Highton, Civil Engineer, of Clarence Villa, Gloucester Road, Regent's Park, London, for Improvements in Electric Telegraphs. Dated September 21, 1850.
>
> My present improvement relates to the manner of protecting

and using insulated telegraphic wires.

The wires for electro-telegraphic purposes, when buried in the ground, or through the sea or rivers, or attached to the walls of tunnels, &c., have generally been insulated with gutta percha, caoutchouc, shellac, pitch, and tar, or other resinous substances, and covered with a leaden or other flexible metallic tube, or have been placed in an iron tube or porcelain tube; such leaden flexible tube, or iron tube, or porcelain tube, being principally used for the purpose of protecting the insulated wires from mechanical injury.

My improvement consists in surrounding the insulated wires or strands of wire, by putting them in the middle of a wire-rope, so that the insulated wires may be surrounded with a flexible covering of iron, or galvanized iron or brass, or other hard wire, or small rods of such materials.

In most cases it is usual in making a wire-rope to place a hempen core in the middle, round which the wires or strands of wires are made to run in spiral curves.

If, then, instead of such hempen core, a wire or wires properly insulated by gutta percha or other insulating material, and, if desired, covered also with a leaden tube, or other flexible metallic tube, or wound round with hemp, coir, or rope, be used instead of such hempen core,—then when the wires forming the wire-rope are twisted round such insulated wires, a wire-rope will be formed, having in the middle for a core one or more insulated wires, as the case may be, and these insulated wires will thus be protected from mechanical injury by such coating of wire-rope or outer wires, and considerable flexibility will also be attained, and the insulated wires will thus be made better able to resist also any lateral or longitudinal strains to which they may be subjected.

I do not deem it necessary to describe the modes of making wire-ropes, as they are now so well known, nor of enclosing therein a hempen core. My improvement, it will be observed, consists in the substitution of an insulated wire or wires for the hempen core usually employed in the manufacture of wire ropes, so that the same may be used for electro-telegraphic purposes, and in the employment, during the manufacture of such wire-rope, of such central core of insulated wires.

Instead of the wire-rope being made circular, a flat wire rope or band may be used, and then by doubling the same over the insulated wires, so as to enclose them therein; and by fastening the rope so doubled over in such position by hoops, bands, or ties, or other convenient means, another form of flexible and strong covering may be given to insulated wires; and possessing also this advantage, viz., by removing the bands or ties, the insulated wires may at any time be exposed, and any repairs done to them.

Such insulated wires, when so protected from external injury, may be used for laying down through the sea or rivers, or under the earth, or for attaching to the sides of walls, or they may be suspended in the air.

In all the above cases, I prefer saturating the rope with a mixture of pitch and tar, or marine glue, or such like substance, either during its manufacture or after it is completed, or at both stages of its condition.

Another patent for further improvements in electric telegraphs has been taken out by the author, the Specification of which falls due in July, 1852.

NOTT'S TELEGRAPH

On the 20th of January, 1846, Mr. John Nott took out a patent for a particular description of an electric telegraph.

In this instrument an electro-magnet of iron acting on an armature causes a pallet (which is so set as to catch into the teeth of a wheel) to force the wheel forward one tooth on the sending of each current of electricity. By the sending of currents of electricity at small intervals of time, the wheel and pointer attached to it may thus be worked round to any desired point on a dial.

Letters were engraved on this dial, and thus any letter might be pointed out by the hand being allowed to rest at such letter for a short period of time.

This telegraph was tried for five miles on the London and North-Western Railway. The old Electric Telegraph Company brought an action against the owners of the patent for a supposed infringement of some of their patents. The matter, however, was compromised by the Company purchasing certain shares in the patent, and taking into their own establishment certain parties connected with the patent.

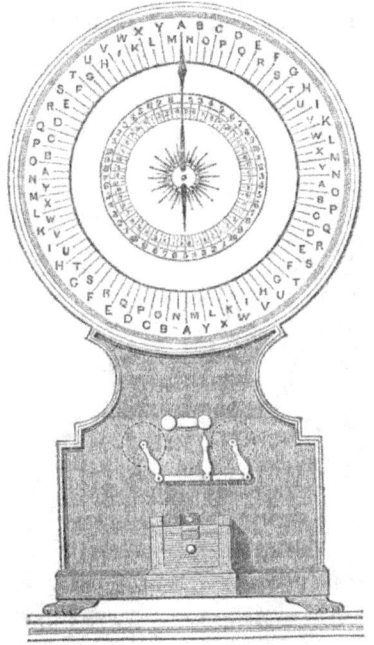 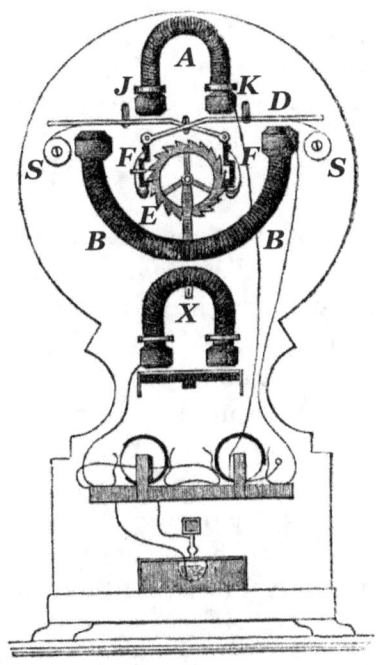

Fig. 37
Nott telegraph instrument

Fig. 38
Nott telegraph (interior view)

The proceedings before the Court of Chancery were very interesting to parties conversant with electricity, and numerous affidavits by men of the highest standing in science were made on the occasion.

This telegraph, although at work for some time between Northampton and Blisworth, is now no longer in use.

Figure 37 above shows the front of one of Nott's instruments. Figure 38 shows the interior.

A and *B* are electro-magnets with armatures *C* and *D*, working on centres *J* and *K*; *E* is a ratchet wheel in which pallets *F* and *F* work.

In this ratchet-wheel the hand shown on the dial in Figure 37 is attached. As the armatures *C* and *D* are attracted to the electro-magnets *A* and *B*, the wheel *E* is forced forward one tooth, and the hand progresses from one letter to the next. A similar movement occurs when the current ceases, the armatures being forced back by the springs *S* and *S*. In this way the hand may be brought successively opposite to any desired letters. *X* is an electro-magnet for sounding the alarum before a communication is made.

POOL'S PATENT

A patent was granted on the 14th of December, 1846, to Moses Poole. Who the real inventor was does not appear, as the patent was taken out in this country as a communication from a foreigner abroad.

The object of this patent seems to have been twofold—the one to improve on the many *detailed parts* of an electric telegraph, and the other to obtain telegraphs free from the patents of Messrs. Cooke and Wheatstone.

The Specification is very long and tedious, and any description here of *each* part would entail the necessity of very lengthy remarks, and would far exceed the limits allowed in this treatise.

There is one novel point, however, which is worthy of remark, and that is, that the metal nickel is proposed as a substitute for soft iron in electro-magnets. The author had himself, a short time after this patent was taken out, the particulars of which were unknown to him at the time, found from a very extensive set of experiments, conducted for the London and North Western Railway Company on various metals, that nickel might be so employed, and he gave instructions for a patent to be taken out for the same in this kingdom; but on learning that a patent had already been secured for that purpose, he at once arranged for the sole use thereof in Great Britain.

The metal nickel has now long been used by the author in his telegraphs on the London and North Western Railway, and with considerable success.

The patent refers also to improvements in insulators, in wires, in magnets, in keys for transmitting currents of electricity, in dials, in magneto-machines,—in short, to most of the detailed parts of an electric telegraph, — but no new principle seemed to be contained in this very long Specification.

Brett & Little's Telegraph

This patent was taken out by Alfred Brett and George Little, and is dated 11th February, 1847. The Specification is a very long one.

In telegraphs under this patent a flat metallic partially magnetized ring was used instead of the magnetic needle. A suitable coil was attached, by which the magnetic ring could be moved to the right or left.

This ring, moving to the right, acted on one pointer, while its motion to the left acted on another pointer; by the single and combined motions of these two pointers the desired letters and symbols were denoted.

The Specification describes many forms of this arrangement, and also a new kind of key or handle apparatus for transmitting the necessary currents of electricity. An alarum was also set free by the motion of this ring.

The Specification also describes an arrangement of lightning conductor,—a mode of insulating wires,—a new arrangement of galvanic battery, in which the exhausted liquid was allowed to run out from the cells and fresh liquid to drop in. An arrangement for the adoption of the magnetic ring to electric clocks is also given.

Instruments under this patent have been used in England for railway purposes.

The old Electric Telegraph Company, conceiving that the way in which the telegraphs of Messrs. Brett and Little were put up were infringements of their plans, took legal proceedings against the patentees. On certain points an adverse decision was come to as regards the defendants, and arrangements were afterwards made with the Telegraph Company.

Henley & Forster's Telegraph

In 1848 a patent was taken out by Messrs. Henley and Forster, for the following kind of electric telegraph:—

Between the poles of an electro-magnet a magnetic needle is placed, moveable on an axle; to this axle a pointer is affixed; a stop is placed, so that the magnetic needle has motion only on one

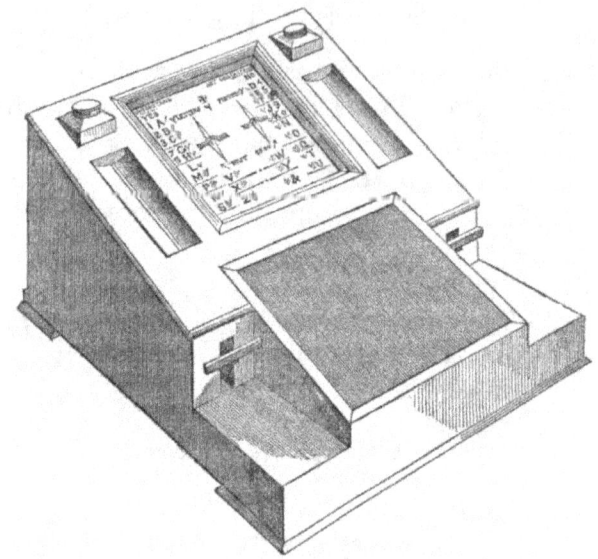

Fig. 39
Henley & Forster Magnetic Telegraph Instrument

Fig. 40
Henley & Forster Magnetic Telegraph (interior view)

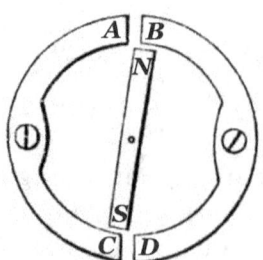

Fig. 41
Magnetic needle and
electro-magnet

side from its normal point of rest. When a current of electricity in one direction is sent round the electro-magnet, the magnetic needle is attracted thereby, and moves from its normal stop to another placed to limit its motion. Instead of the needle returning to its stop when the current ceases, the inductive influence of the magnetic needle on the electro-magnet causes the magnetic needle to remain in that new position until a current of electricity is sent in the contrary direction; and then the magnetic needle is moved from that position to its original position of rest. It thus remains at that stop till another current in the original direction is transmitted. By having two line-wires, and two magnetic needles and electro-magnets, and two pointers, the combination of the motions of the pointers represents the letters of the alphabet.

The electricity produced is from the magneto-machine, as was used by Steinheil in 1837.

An Act of Parliament has been obtained to incorporate a company under the name of the "Magnetic Telegraph Company," for the purpose of working this form of telegraph.

Figures 39 and 40 will explain the arrangement:—

In Figure 41, *N S* represents the magnetic needle, and *A C* and *B D* the horns of the electro-magnet. On pressing down the lever the ends of the armature change place with respect to the poles of the magnet. This produces a current of electricity in the armature, and through the circuit, which, passing round the wire on the electro-magnet, causes it to become magnetic, and thus moves the magnet *N S*, say into the position shown in the diagram. While the handle is kept down, although no electricity is passing, the needle is kept deflected by the induced magnetism in the horns. On allowing the lever to return by the force of the spring on the base, the ends of the armatures and magnets again change places, and a current of electricity is produced in the opposite direction, which entirely neutralizes the induced magnetism, and then reverses the poles of the electro-magnet, bringing the needle to the opposite side.

The author is sorry that the limits allowed forbid his describing the several inventions of Messrs. Barlow, Bakewell, Jacob Brett,

Brown, Clarke, Dering, Dugardin, Hatcher, Little, Mapple, Pulvernachi, Ricardo, Reid, Ward, Williams, and others.

Several patents have also lately been applied for; the particulars of these inventions, however, are as yet unknown.

For a description of the improvements of these parties, some of which relate only to one part of a telegraph, such as the wires for instance, the author must refer those who wish for more detailed information thereon to the Specifications themselves, as enrolled in the Court of Chancery.

An abstract, more or less complete, of some of them, may be found in various periodicals about the dates of those patents.

With the above brief description of the principal features of a few of the patents taken out for *improvements* in electric telegraphs, it is proposed to leave this part of the subject.

It must be understood that although throughout the whole of that description the word *improvements* has been frequently used, it is merely employed in the *patent sense* of the word; and this often means nothing more than another mode of accomplishing a certain known end than that which had been previously proposed or carried out by others.

8
CURRENT TRANSMISSION AND ATMOSPHERIC DISTURBANCES

LAW OF THE TRANSMISSION OF THE ELECTRIC CURRENT

The following mathematical formula expresses the force developed in every part of the circuit by the agency of the galvanic battery: viz:

$$F = \frac{nE}{\frac{nDR}{S} + \frac{lr}{s}}$$

Where n = number of plates:
E = electro-motive force:
D = distance of plates:
R = specific resistance of fluid:
S = area of plates:
s = section of wire:
l = length of circuit:
r = specific resistance of wire:
and F = the force developed, or which is the same thing, its equivalent – viz. the resistance offered to the passage of the electric current.

It is evident, therefore, that in order to obtain a maximum amount of work with the least possible expenditure of power, the coils of the instruments, the number of cells, the distance of the plates, and the resistance of the fluid, ought, on every line of telegraph, to be suited to each other and to the length of the circuit.

Every form and arrangement of battery has a resistance and power peculiar to itself, and as the relation of that force to the resistance of a certain standard unit of wire can only be obtained by direct experiment, and as one person prefers for a telegraph one kind of battery and another person another kind, it would be hardly worthwhile to give here a detailed account of the experiments made by the author on certain particular batteries. It may be well, however, to notice, that in the measurement and comparison of resistances, the author has always used, and was the first to use, as a standard unit of resistance, a mile of No. 8 galvanized iron wire, this being the same as that almost invariably used for telegraphs when the wires are suspended in the air.

Great facilities are afforded by this standard unit of resistance, in adjusting the batteries and coils to the resistance of the line-wire, and to the number of coils included in the circuit.

The unit used is termed "A Mile of Resistance." Thus many of the coils of the instruments of Messrs. Cooke and Wheatstone were found to have a resistance to the passage of the electric power equivalent to twenty-one miles. By using the horse-shoe magnet and coil, as patented by the author in 1848, this resistance in the circuit was reduced in one instance, where four stations were in the circuit, to four miles, and in others to two and a half miles.

The resistance however in the gold leaf of the Rev. H. Highton has been variously calculated at from one-third to one mile only.

To those well conversant with the practical working of the telegraph, where bad insulation is unavoidable, the importance of having a small resistance in the channel laid for the current is too well known to need further remark.

Atmospheric Disturbances of the Electric Telegraph, with a Description of the Means of Preventing Damage to the Instruments

One of the great disturbers of the electric telegraph is to be found in the action of lightning. More damage is often done to the telegraph in a second by a single thunder-storm, than by all the mischievous acts of malicious persons in a whole year. Not only are the magnets demagnetized, but even the coils of wire in the instruments are often fused. In some districts, strips of brass a quarter of an inch wide have been melted by its action.

Current Transmission and Atmospheric Disturbances

Indeed, it is but natural to suppose that if the small currents of electricity produced by the galvanic battery produce so marvellous an effect on a magnetic body when surrounded by coils of wire at a distant station, those powerful discharges of the electric fluid from the clouds, which in their descent burst asunder the steeples of churches, and set houses, trees, and haystacks on fire, should produce disastrous effects on the delicate instruments of the electric telegraph, when the charge happens to be intercepted by the wires of the telegraph, and to be conveyed by them to the telegraphic instruments at a station.

An extract from a Paper "On the Effects of Atmospheric Electricity, as exemplified in the storms of the summer of 1846," written by the author of this book, and read before the Society of Arts in 1846, will show the disastrous effects of lightning on the electric telegraph.

> Having thus far given in detail the course, action, and effects of several heavy discharges of atmospheric electricity during the past summer, I will now proceed with observations on the effects produced on one of the telegraphs under my charge, viz. from Wolverton to Peterborough, a distance of nearly 60 miles. In this case, the instances of disturbance and damage have been too numerous to particularize. Suffice it to say, that I have had the posts struck by lightning and split in twain. I have had parts of the instruments in the stations, and especially the fine coils of wire, melted by the lightning. I have had the needles demagnetized, or their polarity reversed—the bells of the telegraph have been rung by the lightning, and permanent magnetism produced in the electro-magnets; and all this probably in a few hours, and at many stations also at the same moment. Now, however, damage by lightning to the telegraph we fear not. Acting on the principle of the division of the charge, and the peculiarities of high-tension electricity as compared with low-tension, we are enabled to extract from the wires nearly all the lightning before it enters an instrument or a station, and to send it quietly to the earth, and, in fact, to deal with it as we would with the water from the roof of a house, viz. to conduct it where we will. Now, we no longer fear it; for, under the practised hand of science, the lightning itself becomes as tractable as the

power of gravity.

Another example of the action of lightning, that I had the opportunity of examining in the last summer, was a discharge at the Oundle station of the Peterborough branch. The lightning in this case seems to have divided itself over three paths. A portion passed by the leaden gutters of the roof of the station to a tank of water, and thence, by the leaden pipes, to the ground. The tank was of wood, and lined with lead. Immediately underneath it was a gas-pipe, supported by a hook, which was driven into the wood of the tank. The end of this hook all but touched the lead of the tank. The lightning seems to have passed from the tank to this hook, and thence to the gas-pipe and by it to the earth. By the side of this hook was a small air-

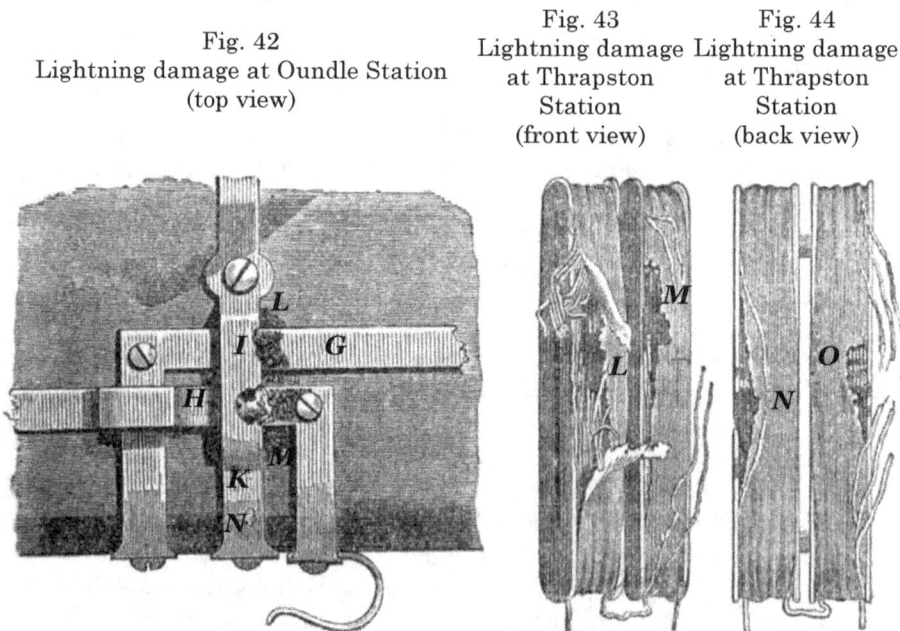

Fig. 42
Lightning damage at Oundle Station
(top view)

Fig. 43
Lightning damage at Thrapston Station
(front view)

Fig. 44
Lightning damage at Thrapston Station
(back view)

hole in the gas-pipe, out of which there was at the time a slight escape of gas. The lightning ignited this jet of gas, and the bottom of the wooden tank was consequently set on fire. If the fire had happened in another part of the station, and had not been discovered in time, serious consequences might have followed.

Current Transmission and Atmospheric Disturbances

Another portion of the lightning passed through the telegraph instrument at the station, fusing some of the metal work therein. It thence proceeded by the telegraph wires to the ground at the next station, Thrapston, a distance of more than eight miles. At this station also considerable damage was done to the telegraph instrument; several of the wires, and some of the metal-work, were fused (see Figures. 42, 43 and 44).

Figure 42. Top view of part of the telegraph apparatus at the Oundle Station of the London and North Western Railway. The strips of brass *G* and *H* were in metallic communication with the wires on the line. The strip *K* was in communication with the ground at Oundle. The strips *G* and *H* were separated from *K* by an interval of about $1/40$ th of an inch. A flash of lightning was intercepted by the wires on the line, and conveyed to this point; but, although the strips *G* and *H* had metallic communication with the earth at Thrapston and Peterborough, yet the resistance offered to the discharge along these directions was such as to cause a large portion of the electric fluid to shoot through the interval between *G K* and *H K*, and to fuse the metals, and produce the effects shown at *G*, *H*, *I*, and *K*. The upper bridge-strip *I*, *K* and the portion of *H* under it have both been melted, and are now firmly united together by the molten metal. The strip *G* had its surface fused, and the strip *I* was melted also. The wood is scorched from *L* to *M*. There is also a melted spot at *N*, on which another portion of the apparatus rested.

Figure 43. Front view of one of the coils of the telegraph instrument at the Thrapston Station of the London and North Western Railway. This coil was burnt and fused on the 1st of August, 1846, by the same flash of lightning which damaged the apparatus shown in Figure 40, although it was more than eight miles distant therefrom! The lightning was conveyed along the wires of the telegraph. The small wires in this coil were fused together, and the silk and cotton burnt off, as shown at *L* and *M*.

Figure 44. Back view of the other coil in the telegraph instrument at the Thrapston Station. Damage similar to that in Figure 41 will be observed at *N* and *O*. The fine wires were all melted together, and the silk and cotton burnt off

In this case, then, we can clearly trace many of the properties of lightning, viz. the division of the charge,—the firing of a mixture of hydrogen and oxygen gases, the fusing of small pieces of metals in its course, and the demagnetization of magnetic needles.

These disastrous effects are now entirely obviated. The plan pursued by the author is as follows: A portion of the wire circuit, say for six or eight inches, is enveloped in bibulous paper or silk, and a mass of metallic filings in connexion with the earth is made to surround such covering. This arrangement is placed on each side of a telegraph instrument at a station. When a flash of lightning happens to be intercepted by the wires of the telegraph, the myriads of infinitesimally fine points of metal in the filings surrounding the wire at a station, and having connexion with the earth, at once draw off nearly the whole charge of lightning, and carry it safely to the earth. This arrangement at once prevents any damage to the telegraph instrument. Not a coil under the author's charge has been fused where this plan has been adopted. The cheapest method is as follows: Line a small deal box, say six or twelve inches long, with a tin plate, and put this plate in connexion with the earth; fill this box with iron filings, and then surround the wire (before it enters a telegraph instrument) with bibulous or blotting paper, as it runs through the centre of the box. All high-tension electricity collected by the wires will at once dart through the air in the bibulous paper to the myriads of points in the iron filings, and thence direct to the earth, and thus the telegraph instrument will be rendered incapable of being damaged even during the most fearful thunder-storms that may occur.

9
REQUIREMENTS FOR AN ELECTRIC TELEGRAPH

The first thing requisite for an electric telegraph is to have a conductor of electricity stretching from one station to another, between which communications are desired to be transmitted. Such conductor should be of metal, inasmuch as metals are the best conductors of electricity.

Although copper is a much better conductor than iron, yet when the wires are suspended in the air, iron wire is now almost universally employed in preference to copper, owing to its possessing far greater strength than copper, and being at the same time much cheaper. The iron wire is generally coated with zinc in order to prevent its rusting.

When rain first falls on the zinc covering, an oxide of zinc is formed, and this oxide being insoluble in water, a second fall of rain cannot dissolve or penetrate it. The zinc covering and the iron wire inside are thus prevented from rusting away. Where the distance between the supports for the wire is very great, as in the crossing of broad rivers, steel wire is employed instead of iron.

The next requirement is, that the wire shall be effectually insulated from the earth, except only at its two extremes. At each of these extremes the wire is metallically united to a mass of metal or coke which is buried in the earth, so that the current may be conducted into the earth at those extremes, and then, by traversing the earth, complete the circuit.

At every point between these two extremes the wire must be well insulated from the earth. For such purpose it is necessary to interpose between the wire and the earth a non-conducting substance. Glass, porcelain, gutta percha, or other resinous substances, are usually employed.

Parties differ in opinion as to which substance is the best. Some prefer porcelain, and others gutta percha. Where glass tubes have been used for the wires to pass through, it is found that strong shocks of electricity crack the tubes of glass just as if they had been cut with a diamond. They are now, therefore, seldom or never used, and porcelain and gutta percha are alone employed.

Telegraphists differ also as to the best form to be employed when these materials are used. Some employ them in the form of tubes, and others in the shape of umbrellas. Each plan has its merits and its defects; the one form being better for some kinds of weather, and the other for other kinds. The author prefers a combination of both; but this is not requisite, unless the line of telegraph is an extremely long one. Either plan will answer for short lengths. Where very excellent insulation is required, the author prefers surrounding the wire with insulating material for a considerable distance on each side of every support. Where iron wires, covered with zinc (being technically termed *galvanized*), are in the neighbourhood of large towns where great quantities of coal are daily burned, the sulphurous vapours arising from such fuel, and passing over the covering of oxide of zinc formed on the wires, convert such *oxide* into a *sulphate* of zinc when the same is covered with moisture. This sulphate of zinc being soluble in water is immediately melted by the rain, and drops off with it. The wire is thus denuded of its insoluble covering, and soon melts away.

The author has had galvanized iron wires reduced from a diameter of an eighth of an inch down to the diameter of a common sewing needle in less than two years.

In such cases it is necessary to protect the wires by a covering of varnish or paint, in order to prevent the contact of these vapours with the wire, or to employ wires entirely encased in gutta percha.

The wires being thus insulated from end to end, the next requirement is a good galvanic battery, or other means of producing electricity. When galvanic batteries are used, the best kind to employ depends entirely upon the nature of the work to be performed by its power.

If it has only to move a magnetic needle surrounded by a coil

Requirements for an Electric Telegraph

of wire, the simplest arrangement of battery is required, such as one composed of cells containing plates of copper and zinc immersed in a solution of sulphuric acid or of the sulphate of alumina. If, on the other hand, the current is required to perform powerful mechanical effects at a distant station, galvanic batteries of a stronger power are necessary,—such, for instance, as Daniell's or Bunsen's batteries. The former are generally used in England, and the latter, viz. Daniell's or Bunsen's, on the Continent. This difference is needed owing to the different classes of telegraphic instruments employed.

KEYS

The next requirement is a key or keys for sending the currents of electricity as desired. The mechanical arrangements of keys are very various. Some persons prefer them on one plan, and some on another. Where currents of electricity are required to be sent either positively or negatively, the author prefers for ease, rapidity, and simplicity of mechanism, the following plan, as shown in Figures 45 and 46 overleaf.

A and B are two keys of ebony or ivory working up and down by means of the spring-joints C D and E F.

The two spring-joints C and E are metallically united to one pole of the battery, and D and F to the other pole. These spring-joints are prolonged under the keys A and B, and are connected respectively with the upright screw-pins or studs, G and H and I and K.

Underneath these keys is a moveable spring, L, pressing upwards and resting against the stud M.

When the spring L is pressed away from the stud M, the spring L and stud M are insulated from each other.

The severed ends of the telegraphic wire are to be metallically united with L and M.

The action of the keys is as follows:

On pressing down the key A, the spring L is thereby forced by the stud H away from the stud M, and the stud G then comes into metallic contact with M, while the stud H remains in contact with L. A current of electricity in one direction would thus pass over

Fig. 45
Highton's telegraph keys (side view)

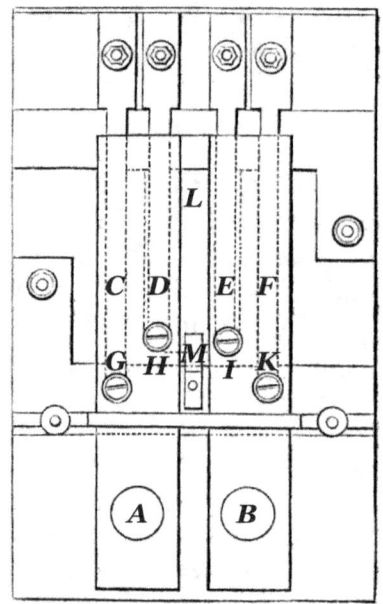

Fig. 46
Highton's telegraph keys (top view)

the line-wire. On pressing down the key **B** a current in an opposite direction would be similarly sent, as is evident.

It will be observed that in the above keys the making and breaking of contact for sending either a positive or a negative current of electricity requires the use of only one moveable spring.

It will also be observed that this arrangement of keys gives a maximum of simplicity, for there cannot be less than one moveable spring employed, and every additional spring used would not only render the mechanism more complicated, but much more liable to derangement. On the perfect action of every spring in the electric circuit depends the working of the telegraph. If one spring fail or become inactive, the whole telegraph from end to end is deranged. It is most desirable, therefore, that the number of springs in the circuit should be as few as possible.

ALARUMS

We will now pass on to a description of the alarum to be employed for calling the attention of the distant attendant. Before sending a telegraphic communication, it is desirable to ring a bell for calling the attention of the attendant to the telegraphic instrument, especially where such person has other duties to perform than those of attending to the telegraph.

This is accomplished by placing an electro-magnet, surrounded by coils of wire, in the circuit at such station. An armature is suspended in close proximity to the electro-magnet. This armature is attracted to the electro-magnet when a current of electricity is sent along the line-wire.

By the motion of such armature, wound-up mechanism is liberated, and thereby allows a hammer to strike a bell. This calls the attention of the attendant to his telegraphic instrument. He then sends back a current of electricity, and thereby similarly rings a bell at the station at which the person is who desires to send the communication. This is to show that all is ready for receiving the message.

Both parties then, by moving a sliding-bar of metal, remove from the circuit electro-magnets of their respective alarums, and leave only in the circuit the keys and the telegraphic instruments for sending and receiving the communication.

The forms of alarums used for such purpose differ much in mechanical detail. It is thought well, however, to give examples of two kinds only, viz. those of the latest invention: these are on two of the several plans patented by the author, and belonging to the British Electric Telegraph Company.

The first, Highton's Alarum No. 1 (see Figures 47 and 48), is one which requires but a small amount of electricity for liberating the mechanism, and is at the same time very perfect and certain in its action.

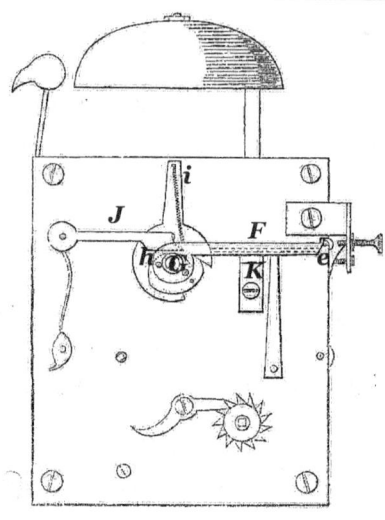
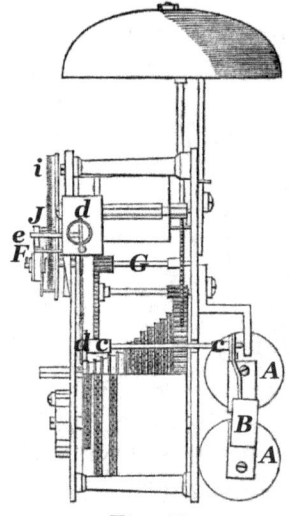

Fig. 47
Highton Alarum No. 1
(view inside the case)

Fig. 48
Highton Alarum No. 1
(side view)

DESCRIPTION OF ALARUM NO. 1

An electro-magnet **A A** has an armature **B** working on an axle **c c**, to the other end of which is fixed an arm **d d** with projecting catch **e**. The arm **F** on axle **G** rests against the catch **e**, pressing with a very slight power, which is caused by the spring **i** fixed to the eccentric **h**, which works on the axle **G**. The detent **J** of the wound-up mechanism rests on the eccentric **h**. On an electric current being sent round the electro-magnet, the armature **B** and catch-lever **d e** are moved. This allows the arm **F** to rotate upon its axle **G**, the motive power being the force of the spring **i**, which is so set that the power, before the arm is released, is only just sufficient to start it; the spring **i**, acting on the eccentric principle, increases its power as the eccentric rotates, and raises the detent **J**, and thus liberates the mechanism of the alarum, causing the hammer to strike against the bell. As the works of the alarum rotate, and with it the arm **F**, the end of the arm **F** is left under the catch **e** (which, on the cessation of the current, returns to its place of rest). The spring and eccentric are stopped when in their former position, and there detained; thus making the alarum ready for a second action.

Requirements for an Electric Telegraph

K is an arm which, at the required time, is pressed out by a projection or cam on the eccentric plate, so as to meet the arm **F**, and break the force of its blow when returning to its place of rest against the catch **e**. Thus the arm **F**, on returning to the catch **e**, has its force broken, and has a momentum due only to its motion from the point where **K** catches it, to the point where it is stopped by the catch **e**.

The next plan, Highton's Alarum No. 2 (see Figures 49 and 50), is used where it is thought desirable not to remove the alarums at all from the circuit, lest the attendants should neglect to replace them in the circuit when the communication is finished.

This is accomplished by causing the currents of electricity of *long duration* to sound the alarum, but currents of electricity of *short duration* to be unable to do so. Thus currents of electricity of long duration sound the alarum, whereas currents of short duration are caused to make the signals necessary for the communication. This plan is quite new, and was only patented on the 28th of January, 1852.

It is believed that the adoption of this principle will he found of great utility in telegraphs for railway companies where many stations are embraced in the same circuit, for otherwise each attendant at every station in that circuit has to remove his alarum from the circuit before a message is transmitted between any one station and any other in that circuit, and to replace the same after such message is finished, unless the alarums are allowed to continue ringing during the whole period of the transmission of the communication, or unless an entirely distinct wire is set apart for the alarums.

DESCRIPTION OF ALARUM NO. 2

A A is an electro-magnet with armature **B** fixed on an axle **E** working on centres **c c**. **D** is an arm fixed to the axle **E** and having upon it a catch **F**. **G** is a wheel connected with the wound-up machinery. This wheel **G** is kept in a state of continual revolution. **H** is a small axle fixed eccentrically upon the wheel **G**. Upon this axle **H** an arm or lever **i l** is placed with a weight or hammer **K** at the end of it. This arm **i l** is prolonged to **l**, so that when the electro-

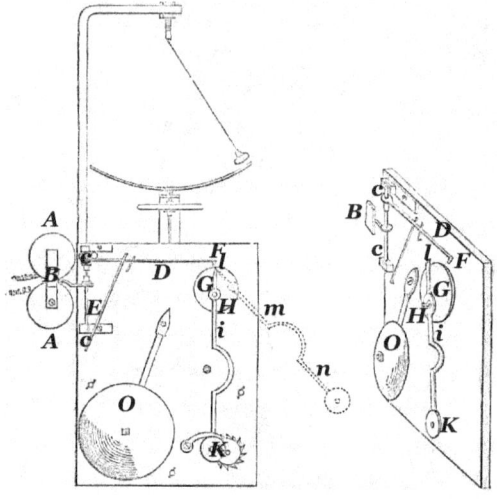

Fig. 49
Highton Alarum No. 2
(view inside the case)

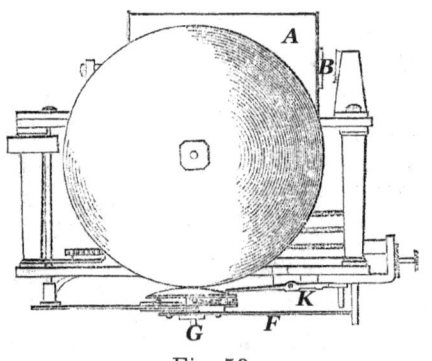

Fig. 50
Highton Alarum No. 2
(top view)

magnet **A A** attracts the armature **B**, the catch **F** would be interposed in the way of the end *l* of the arm *i l*; but when the armature is not attracted to the electro-magnet, the catch **F** does not interfere with the free motion of the arm *i l*.

The motion of the wheel **G** is so regulated by the motion of the wound-up mechanism that it is caused to revolve once in a given definite period of time,—say one, two, three, or more or less seconds. When a current of electricity is caused to pass round the electro-magnet *A A* and to continue for a long period, the end *l* of the arm *i l* is intercepted in its free motion in each revolution by the catch *F*. The arm *i l* is then gradually thrown into an oblique position until it arrives in the position shown by the dotted lines *m n*, and it then escapes from the catch *F*. The hammer *K* being then at liberty to resume its wonted position, seeks to do so by the action of gravity or of a spring, and by the accumulated momentum obtained in doing so, it passes by such wonted position and strikes against the bell *O*. Now if the current be not continued for a long period, but only for periods of short duration, such as are used in sending ordinary telegraphic communications, then the catch *F* being suddenly released from the end of the arm *i l* before the arm *i l* is thrown into the position shown by the dotted line *m n*, the weight *K* in resuming its wonted position does not acquire sufficient momentum to carry it over to the bell *O*. In this case the bell will not be struck, but in the former case it will.

This arrangement therefore of an alarum will render it possible to continue the electro-magnets of the alarums in the *same circuit* with the telegraphic instruments *without* ringing the bells, and without the necessity of removing the coils of the electro-magnets from the circuit when a telegraphic communication is being transmitted, or requiring another distinct wire for the alarums. But, nevertheless, the alarums may be sounded, whenever desired, by merely continuing the currents of electricity for a longer period than that required for sending telegraphic communications by the indicator.

INDICATING INSTRUMENTS

We will pass on now to the telegraphic instruments employed for reading off the communications transmitted. One of the simplest forms is that in which a magnet fastened to a moveable axle is acted upon by a coil of wire which forms part of the electric circuit. To this axle an indicator of any desired kind is affixed, which moves to the right or left along with the magnet, accordingly as the current is sent positively or negatively.

By the number of these motions to the right or left, and of

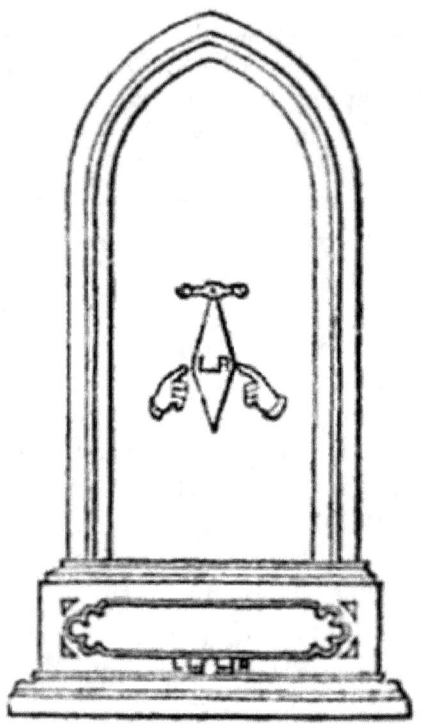

Fig. 51
Highton Disc Telegraph

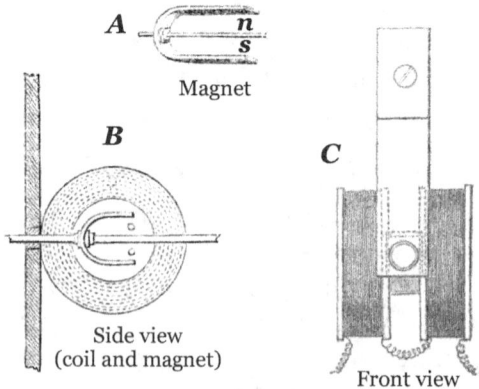

Fig. 52
Highton Disc Telegraph (internal features)

their combinations, the letters of the alphabet are denoted.

Different telegraphists, although they employ the number of such motions to represent the letters, do not all use the same number of motions and of combinations of motions to designate the same letters, but many employ different codes. The principle, however, is the same, and is one of the oldest plans used.

Some parties use two needles with two wires, and employ the number of single and combined angular movements of such two magnets to denote the letters.

Two wires, however, if used for two separate single indicator instruments, will, it is found in practice, do more work than two wires combined with two magnets in one and the same instrument.

Figure 51 represents the most modern form of telegraphic instrument requiring but one line-wire, and having a disc with two letters thereon. The disc is moveable to the right or left, and when so moved, the stationary hands or indicators point out the letters thereon which are intended to be denoted. By a repetition of these movements the alphabet is formed.

In the above instrument the simple arrangement of keys before described is inserted.

The magnet is of the horse-shoe form, and the axle passes through the middle of the magnet, and is parallel to the limbs thereof (see Figure 52, *A*,).

The magnet is enclosed in a coil which forms part of the electric circuit (see Figure 51, *B* & *C*).

In this arrangement of magnet and coil the electricity acts on *both* poles of the magnet at the same time.

When it is required to send messages of great length, it is desirable to use a plan of telegraphic instrument which *records* the communication *on paper*; or otherwise the eye would soon become wearied by watching the rapid movements of the disc, and mistakes would inevitably take place.

Various plans have been proposed for such purpose, and several different arrangements are now in use.

The simplest method appears to be that of causing an electro-magnet to stamp or cut a piece of paper, and to make thereon long or short marks accordingly as the current is continued for a long or short period.

This is the plan almost universally adopted in America and on the Continent, but it has not as yet been practically employed in England.

The British Electric Telegraph Company are about to employ a modification of this arrangement as the basis for the plans to be used for their heavy commercial work.

They intend, for some kinds of intelligence, to use an arrangement by means of which a positive current of electricity will make one mark on paper, and a negative current a different and dissimilar mark, according to the patented plans which they possess.

In all cases they will use a secondary power for the purpose of so marking the paper, in order to supersede the necessity of the transmission of powerful electric currents along their line-wires.

In this way far less marks are required for forming letters or words.

INSULATION

In the construction of an electric telegraph the greatest possible care and attention should be paid to the insulation of the wires.

It is impossible to insulate the wires too well. Defective insulation is a source of the greatest annoyance and disappointment. However accurately in mechanical detail the telegraphic instruments at the stations are made, it is impossible that they can work well if the insulation is imperfect.

Much, it is true, may be done by increasing the power of the galvanic batteries, but all this additional power is in itself an utter waste, if it is required to supply the defect of imperfect insulation.

In such case the instrument nearest to the battery will work powerfully, while the distant instrument, if it work at all, will only do so with great sluggishness.

To obtain perfect insulation is impossible, as no substance has yet been discovered which is an absolutely perfect insulator. And again, when wires are suspended in the air, the occurrence of fogs, damp weather, rain, and showers of sleet deprive the wires of a great part of the insulation which they possess in dry weather.

The mode therefore of insulation employed should possess the best antidotes to *all* the variations of weather.

Where many wires are suspended on the same posts, defects arising from these causes are greatly multiplied.

In America, where only one wire is used, injury arising from the above causes is not so detrimental; but in England, where occasionally as many as 20 or more wires are placed on the same post, the action is most detrimental, and electricity, when intended to be transmitted along one wire only, often finds its way more or less into all the wires, and thus not only lessens the quantity intended to be transmitted to the distant instrument, but disarranges the instruments connected with all these other wires.

Owing to this result of imperfect insulation it has been found impossible for weeks together to telegraph direct even between London and Liverpool, although the insulation has been changed several times in the course of the last few years. The only way in which the communication could be carried on was by means of sending the message to an intermediate station, and then by repeating it at such intermediate station, to forward it thence to Liverpool.

The author hopes in the telegraphs he is now erecting to render any repetition unnecessary, by adopting the following plan.

Where many wires are required to be suspended on the same post or support, he proposes, first, to conduct the insulation to a considerable distance from the post, and secondly, to place between wire and wire a direct communication with the earth, so that any of the electricity transmitted, as it escapes from the wire, may be intercepted by this communication with the earth, and so transmitted direct to the earth without the possibility of its entering an adjoining wire.

In such case it will only be necessary in very great lengths of wire and in very adverse weather to increase the quantity of electricity transmitted, in order to make due allowance for the quantity that escapes at the points of the insulation. How ever great such amount of electricity required may be, no portion thereof can reach the adjoining wire and thereby disarrange the telegraphic instruments connected with such other wires.

COILS

With respect to the diameter and length of the wires to be employed in surrounding the electro-magnets, or the magnets to be used in the telegraphic instruments, these must be made to suit the particular length of the circuit and the size of the electro-magnets or permanent magnets employed, as well as the kind of galvanic battery and the number of cells therein intended to be used.

Each telegraph will require a separate calculation, in which the resistance of the circuit and the battery itself will form an element. No precise rule therefore can be given, but the whole will depend on the proper solution of the equation as propounded by Professor Ohm, and given at page 105. The engineer who is acquainted with the laws of this science will find no difficulty in the solution of the problem, whereas to the public, or to those ill-versed in the arcana of the science of electricity, it would be impossible to make the solving of this problem intelligible.

QUANTITY OF ELECTRICITY TO BE USED FOR TELEGRAPHING

Many persons even well conversant with the electric telegraph have considered that the sole end to be obtained was to devise a telegraphic instrument in which the smallest quantity of electricity possible should effectually produce the necessary signals without making due provision for the action of atmospheric electricity. These persons seem to have entirely forgotten that in many states of the atmosphere strong currents of electricity pass between the clouds and the earth, and are intercepted by the extended wires of the telegraph.

It is obvious in the case of the use of instruments thus unprotected, that these natural currents of electricity in their passage to the earth would be intercepted and collected by the telegraphic wires, and operate upon and disarrange the telegraphic instruments connected with such wires.

It is evidently therefore *not* desirable that the telegraphic instruments should be constructed so as to work with an *extremely small quantity of electricity*, unless a perfect means be resorted to, to render harmless the effects of atmospheric electricity, for in such case *every natural* discharge would affect and disarrange the telegraph.

On the other hand it is undesirable to employ a telegraphic instrument which requires the use of a very large amount of electricity. The telegraphic instruments should, therefore, be so constructed that the ordinary discharges of atmospheric electricity, as developed in fogs, or the action of the aurora borealis, should not be enabled to affect the same, nor yet heavy discharges of the electric fluid during the presence of thunder storms; and in this latter case, nearly the whole of the electric fluid intercepted by the wires may, before it reaches the instruments, be extracted therefrom by means of a proper construction of lightning conductor, and the instrument thus made proof against *all* weathers.

Lightning Conductors

One of the best arrangements of a lightning conductor for such purpose consists in wrapping the line-wire, before it enters the telegraphic instrument, with silk or bibulous paper, and surrounding such covering with metallic filings placed in direct metallic communication with the earth.

By such arrangement nearly the whole quantity of electricity intercepted by the wires from a flash of lightning is extracted by the myriads of fine points of the metallic filings surrounding the wire, and conveyed harmlessly into the ground, without in any degree deranging or damaging the coils or magnets of the telegraphic instruments.

When the insulation is conducted upon the principles already described, and the length and size of the wire in the coils and the magnets employed are made in compliance with the above rules, and proper lightning conductors attached near to the instruments, the telegraph so constructed will cause a minimum quantity of electricity to perform a maximum amount of work, and the telegraphic instruments will *at the same time* be rendered incapable of being injuriously acted upon by extraneous currents of electricity traversing the atmosphere through which the wires pass.

SUBTERRANEAN WIRES AS COMPARED WITH WIRES IN THE AIR

It may be said that much of the alleged damage likely to ensue from the action of natural currents of electricity passing through the atmosphere would be obviated by the use of wires buried in the earth. But when it is found, in the case of even a single line of telegraph in Prussia, that more than a hundred miles of wire which were buried in the earth have been abandoned (owing to their defective insulation, and the difficulty experienced and the time occupied in detecting the exact position of those defects, and in remedying them when discovered), that the wires were suspended on posts in their stead. Therefore the employment of subterranean wires for the sake *merely* of lessening the effects of atmospheric electricity cannot be recommended.

And again, when we call to mind the great additional expense that must be incurred at the first outset, and the great difficulty and expense that must be encountered afterwards in burying *additional wires* when the increasing wants of trade demand such additions, it would appear unwise, on the present unsatisfactory evidence on the subject, to pursue very extensively the plan of burying the wires in the earth in preference to their suspension in the air, unless money were of little or no importance, and the best possible insulation was demanded, whatever might be its cost.

ON THE VELOCITY OF THE ELECTRIC CURRENT

The first persons who observed the great velocity of the electric current were Otto Guericke, Gray, and Wheeler. They believed that the velocity was all but infinite.

The celebrated experiments of the Abbé Nollet at Paris demonstrated the same fact. Nollet, in the presence of the French Court, passed a discharge of electricity at Paris through six hundred persons. Every person appeared to receive the shock at the same identical instant.

On the 14th and 18th of July, 1747, Dr. Watson passed a charge of electricity through a wire extending over Westminster Bridge at London, the return circuit being made by the water of the Thames. On the 14th of August, in the same year, Dr. Watson formed a circuit of two miles of wire and two miles of earth, and passed electric currents through this circuit. The period of time occupied by the passage of the power was apparently infinitely small; the throwing in of the charge at the one end, and its reappearance at the other, being sensibly at the same identical instant of time. It was reserved, however, for the illustrious Wheatstone to prove, by mechanical means, what the rate of the transmission of this power really was.

By means of the child's toy, viz. waving a piece of burning stick in the air, it was known that if the velocity were very great, the flaming end of the stick presented the appearance of a ribbon of fire. A piece of burning wood was then fixed to a revolving axle. If the ignited end made a whole revolution in the $1/10$ th of a second, it formed a complete circle or circumference of light; if a less velocity were given to it, the circumference would appear broken. This at once proved that the impression produced by light on the retina of the eye lasted for a period of about the $1/10$ th of a second.

This fact being borne in mind, a bright brass arm was substituted for the ignited piece of wood, and the apparatus placed in the dark. This brass arm was, by means of mechanism, made to revolve with different velocities. A spark from an electrical machine or from a Leyden jar was then produced. The light of this discharge illumined the revolving arm of brass, but the *duration of the spark* was so short that, however fast the arm revolved, it always appeared stationary. Now, if the spark had endured for even the $1/10$ th part of a second, the arm would have been illumined the whole time, and it would thus have been rendered visible to the eye during that period, and have appeared as a plane surface. It was proved, however, by this experiment that the duration of the spark was not even the $1/10000$ th part of a second of time.

Flying insects and the vibrating strings of musical instruments were subjected to the same test, yet each on being illuminated by the electric spark appeared as perfectly stationary. Experiments were also similarly tried on jets or falling drops of water, yet the same result was invariably produced.

It now became an object to ascertain how rapid the passage of electricity really was.

The Electric Telegraph–Its History and Progress

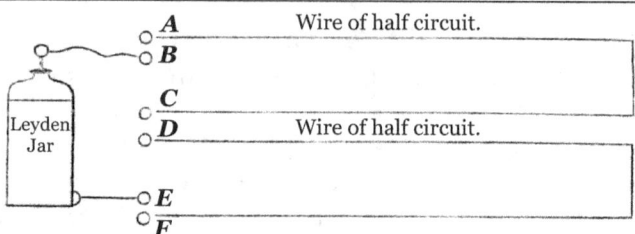

Fig. 53
Experiment to determine the speed of electricity

A long circuit of wire was erected in the Adelaide Gallery, the middle thereof being intersected, and two brass knobs placed at their severed ends, and separated by a distance of about $1/10$ th of an inch. Two brass knobs were placed at the extremes also of this circuit, the one knob being $1/10$ th of an inch apart from a knob in metallic connection with the inside of a Leyden jar, and the other knob being about the same distance from a knob in metallic connection with the exterior coating of the same jar.

The arrangement, therefore, was substantially the same as that shown in Figure 53.

On the discharge of the Leyden jar a spark would pass, as is evident, between *A* and *B*, *C* and *D*, and *E* and *F*.

Now, if these three sparks occurred at apparently the same instant of time, the passage of electricity between the inner and outer coatings would appear to be instantaneous.

In order to ascertain whether this were so or not, mechanical arrangements had to be devised, for the unassisted eye could not decide with absolute certainty.

The following was the mechanical arrangement employed for the purpose. A vertical mirror was fixed to an upright axle which was caused to revolve with very great rapidity in front of the knobs *A B*, *C D,* and *E F*.

Now if the sparks passed between those knobs at identically the same instant of time, all would appear in the revolving mirror in a vertical line, thus: ▮

If, on the other hand, the spark between *A* and *B* preceded that between *C* and *D*, and the spark between *C* and *D* preceded that between **E** and **F**, their reflected images would appear thus if the mirror revolved from right to left: ▮ and thus: ▮ if the mirror revolved from left to right. On trying the experiment, however, it was found that the reflection of the sparks appeared thus: ▮ if the mirror revolved from right to left, and thus: ▮ if it revolved from left to right.

This gave the following results, viz.

First, that the electricity started from each coating of the jar at the *same instant,*

And secondly, that time *was* required for the electricity to travel to the middle of the circuit.

It now became of immense importance to ascertain how long the period of time occupied by the passage of the electricity from one *end* of the circuit to the *middle* really was. On ascertaining the distance of the upper or under reflections with that of the middle reflection as measured horizontally, and comparing that distance with the velocity of the motion of the mirror, the velocity of the electric current was at once determined.

It was thus proved that the current did not travel at a less speed than 200,000 miles a second—a velocity greater than even that of light itself.†

The velocity, therefore, of the electric current was thus proved amply rapid for all possible requirements in the electric telegraph.

Other equally ingenious instruments have been devised by Professor Wheatstone for measuring by means of electricity distances over which bodies travel with great velocity, and that too with such a degree of accuracy as to measure the distance to the $1/1000$ th part of an inch. Thus, for instance, the velocity of a bullet may be measured when it emerges from the muzzle of a gun, as also the time occupied in its transit through the barrel.

In the same way the time that elapses between the striking of the percussion-cap of a gun and the starting of the bullet may

---

† **Editor's Note:** With the benefit of Einstein's theories of relativity, we now know that nothing can exceed the speed of light - approximately 186,000 miles a second.

be ascertained with the greatest accuracy.

The author has himself, by a modification of one of Professor Wheatstone's instruments, measured the time that elapses during the descent of a body in falling through a distance represented by merely the thickness of a sheet of common writing-paper.

As a full description of such instruments will be found in the "Transactions of the Royal Society," it is deemed unnecessary to dwell further on this part of the subject.

10
ELECTRIC TELEGRAPHS NOW IN USE

INVENTION OF THE ELECTRIC TELEGRAPHS NOW IN USE

No *one* person can be strictly called the inventor of the electric telegraph. In order to ascertain whether the honour of this magnificent invention can be ascribed to any *one single person*, let us *dissect* any of the forms of telegraph now in general use.

Let us take, for example, the needle telegraph as now generally used by "The Electric Telegraph Company" in England, and by its dissected parts show to whom is due the honour of the invention or discovery of each of its parts.

- Volta, in 1800, discovered the galvanic current.
- Oersted, in 1811, discovered that a magnetic needle was moved by the passage of an electric current through an adjoining wire.
- Schwieger invented the coil.
- Schilling, in 1832, placed the magnetic needles vertical.
- Steinheil, in 1837, made the counting of the number of motions the basis of his alphabets.
- Sturgeon discovered and invented the electro-magnet.
- Schilling, in 1832, used a weight which was caused to fall by a current of electricity to sound a bell.
- Cooke and Wheatstone, in 1845, made a similar falling weight to liberate wound-up mechanism, and thus to sound a bell.
- Early experimenters showed that glass, porcelain, and resin were insulators of electricity.
- Watson, in 1747, sent currents of electricity through wires suspended in the air on posts.

- Steinheil, in 1837, used wires suspended in the air, and buried in the ground, for an electric telegraph.
- Cooke, in 1842, patented a *particular method* of suspending wires in the air, and a particular *form* of glass and porcelain for insulators.
- Watson, in 1747, showed that one-half of an electric current might be formed of the earth.
- Steinheil, in 1837, used the same for telegraphic purposes, as did also Cooke and Wheatstone in 1842.

No one can then say that any *one* person is the inventor of the electric telegraph, as now generally in use in England; in which galvanic batteries,—coils of wire,—moveable magnets,—electro-magnets,—wires on posts in the air,—wires underground,—earth plates,—and the counting of the signals to compose the alphabet, form the entire telegraph.

Each one of the above-named persons has a right to claim the discovery or invention of *one* of those parts, and numerous others have an equal right to claim the invention of many of the *combinations* contained in such telegraph.

The same may be said of every telegraph now in use throughout the world.

The *complete telegraph* is a joint invention. Each person has employed in building up his particular telegraph the discoveries and inventions of many others.

Too great praise, however, cannot be bestowed on Mr. W. F. Cooke for his unwearied and energetic exertions in putting into practical operation the inventions of Professor Wheatstone and himself.

It is owing to these exertions on the part of Mr. Cooke that this kingdom can now boast of having received the benefits of telegraphic communication as early as any other country.

Prevailing Ignorance as to the Action of the Electric Telegraph

Many persons there are, even at this date, who have not the most remote idea as to the manner in which the electric telegraph works.

Some firmly believe that the paper on which the message is written actually passes through the interior of the wire itself.

Others declare that on placing their ears against the posts, they can hear the communications passing. But none of these latter wiseacres have ever been able to ascertain thereby what the communication is.

The Aeolian sounds produced by the wire when acted on by the wind have led many persons into this belief, and numbers there are, even amongst the educated classes, who believe that the sounds so produced are at any rate the effect of the passage of the electric current.

It is no uncommon thing to hear an illiterate person say, "What lots of messages were sent by the *telegraph* yesterday!"

On questioning him as to the grounds of his knowledge, he unhesitatingly affirms that he knows it well enough, for he *heard* the wires *at it* almost all day.

But perhaps one of the most ludicrous of such like cases is the following:

An old woman, hearing the bell ring at the Station for the people to take their seats in the train, and the guard calling out "Now then, those that are going, get in," suddenly rushed into the carriage, leaving her umbrella in the hurry of the moment on the platform. The guard, as the train started, observing the umbrella, and believing it belonged to someone who had just entered the carriages, put it into his van. He was right in his surmises, for at the next station the said lady, on getting out of the carriage in the rain, missed her umbrella.

"Och, 'faith," says she, "and there's that dear darling umbrella of mine left behind. What shall I do? I would not have it lost for the whole world—it was the gift of my dear old mother—poor cratur'. Och, 'faith, and what *shall* I do?"

The guard hearing the lamentations of the old woman, and wishing to have a bit of fun, whispered to a porter to take the umbrella out of his van and to hang it on the wires; and while he went into the station.

"Be calm, madam," said the guard, "I will telegraph by the

wires, and see if you did leave it."

Off he went to the telegraph instrument, followed by the woman. He rang the bell violently, moved the needles, and immediately shouted "All right, ma'am, you did leave your umbrella as you said, and they have sent it by the telegraph. If you will look outside, you will find it *on the wires.*"

Out ran the old woman, and, to her unbounded astonishment and delight, there her umbrella, sure enough, was on the wires. She seized it in ecstasies of delight, and looking earnestly first at the umbrella and then at the posts, exclaimed in utter astonishment, "And 'faith, and how did the poor cratur' pass by the posts without even a scratch or a mark on it, poor soul ? "

The guard replied, "Ah, ma'am, if I were to explain it to you, you would not understand it. We have queer things in this country."

"And sure enough we have, your honour," replied the old woman, "and if the inventor of that 'ere thing don't go to Heaven, nobody ought;" and off she trudged with her treasure, well satisfied with the performance of the electric telegraph.

From that day to this, this old lady, no doubt, firmly believes that in the twinkling of an eye her dear old darling umbrella, at the sounding of the telegraph bell, passed over some ten miles of wire and some hundreds of posts, and arrived safely at its destination.

With this short digression, it is now proposed to pass on to a very brief description of the various kinds of telegraphs in use in Europe and America.

TELEGRAPHS IN GREAT BRITAIN

In England, the patents of Messrs. Cooke and Wheatstone were bought by The Electric Telegraph Company in 1846. In the same year this Company obtained their Act of Incorporation. They being the first Company in the kingdom, have supplied most of the leading lines with telegraphs upon the principles of their patents.

The kind of instrument now generally used by them is the double-needle telegraph, a drawing of which has already been given (see page 80).

The letters are denoted by counting the number of move-

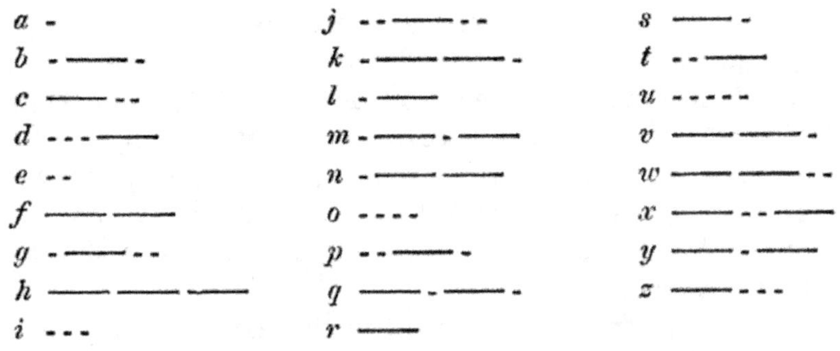

Electric Telegraphs now in Use

```
a  · —              j  · · — · ·        s  — · ·
b  — · · ·          k  — · —            t  —
c  — · — ·          l  · — · ·          u  · · —
d  — · ·            m  — —              v  · · · —
e  ·                n  — ·              w  · — —
f  · · — ·          o  — — —            x  — · · —
g  — — ·            p  · — — ·          y  — · — —
h  · · · ·          q  — — · —          z  — — · ·
i  · ·              r  · — ·
```

Fig. 54
Alphabet code used by The Electric Telegraph Company

ments of the needles to the right or left before a slight pause is made, —a principle which was adopted by many inventors of telegraphs before any patent was taken out.

This Company have long had the exclusive enjoyment of sending all the telegraphic intelligence of this country. They paid no less than £168,000 for their patents.

Another form of telegraph is also used by this Company at a few of their large commercial stations. It is a chemically marking telegraph, in which lines and dots are made on a moving ribbon of paper, by means of currents of electricity. The alphabet used is very similar to that of Professor Morse, in America.

Figure 54 shows the alphabet code now used for that instrument.

The telegraphs of Messrs. Highton have also been used to a considerable extent on the London and North-Western Company's lines of railway, viz.: on the main line; the Peterborough Line; the Liverpool and Manchester Line; the Leeds and Dewsbury Line; and the Manchester and Huddersfield Line.

They are now about to be extensively used throughout the kingdom by the British Electric Telegraph Company, who, as it has been before stated, have bought the patents, and have obtained an Act of Parliament.

The telegraphs of Messrs. Brett and Little have also been used, but having in certain points been proved by a Court of Law to be infringements of the patents of the Electric Telegraph

Company, and a compromise having been lately come to between those two parties, it is probable that the instruments will be removed, and replaced by those of the Electric Telegraph Company.

The great *distinguishing feature* then of the *majority* of the telegraphs at present used in Great Britain is, that they are of the class known as oscillating telegraphs, *i e.* telegraphs in which the letters are denoted by the number of motions to the right or left of a needle or an indicator.

The wires in England have hitherto been almost universally erected on posts. The insulation however is at present by no means perfect.

Extent of Electric Telegraphs in use in Great Britain

According to official returns, there were in April, 1850, 5,447 miles of railway open and at work, 1,784 miles in course of construction, and 4,795 in abeyance; making a total of 72,31 miles of railway opened or opening, and 4,795 miles of railway in suspense.

According to the published returns of the old Electric Telegraph Company, wires were erected over only 2,215 miles of railway, leaving at that period thousands of miles of railway without any telegraph thereon at all, besides thousands of miles of railway for which Acts of Parliament had been obtained, but on which no works had been commenced.

Since that period, considerable progress has taken place in the construction of Electric Telegraphs, but even now many portions of the kingdom are wholly deprived of the benefits resulting from telegraphic communication. The exact length of telegraphs constructed from the above period up to the present time has not been published.

Again, with respect to Ireland, upwards of 500 miles of railway are in operation, but not five miles of telegraph are yet in use. A contract, however, has lately been entered into for the construction of a telegraph on one line in Ireland, viz. between Dublin and Galway.

Electric Telegraphs now in Use

Fig. 55
Mr Jacob Brett's Telegraph

TELEGRAPHS IN AMERICA

In America three kinds of telegraph are in use, viz. Morse's, Bain's, and House's. Morse's telegraph, which has already been described, is at present the most generally employed. Bain's chemically marking telegraph comes next; and, lastly, Professor House's, which is on the step-by-step movement principle, and which prints in ordinary letters the intelligence transmitted.

This telegraph is at present used only to a very limited extent in the United States, and is known in England as the telegraph of Mr. Jacob Brett. It is illustrated in Figure 55.

In point of rapidity, Bain's telegraph would come first, Morse's next, and lastly, House's, which must be slow as compared with either of the other two. Considerable improvements, however, have been lately made in the telegraphs of Messrs. House and Brett.

It will be observed that the general system of telegraph used in America differs entirely from the oscillating telegraphs employed in Great Britain, the telegraphs in America being all on the permanently recording principle.

The wires in America are erected entirely on posts, except at a few crossings of large rivers, and there wires covered with gutta percha are laid under the water.

For a description of Morse's telegraph, see page 64.

EXTENT OF ELECTRIC TELEGRAPHS IN AMERICA

It is notorious how very far in advance of Great Britain America has long been in the means of telegraphic communication, as compared with its commercial position.

There were, in the beginning of 1850, in the United States, no less than 12,000 miles of telegraph in practical operation, and 3,000 miles of telegraph under construction.

In America the telegraph is used to an enormous extent. This is partly owing to the low charges made for its use and the rivalry resulting from competing companies.

On the 1st January, 1848, there were only 2,311 miles of telegraph in operation, whereas on the 1st January, 1850, there were 12,000 miles completed and at work. These 12,000 miles of telegraph were the property of no less than twenty telegraph companies. At the present time there are nearly 30,000 miles of telegraph in operation and under construction in America.

Many of these telegraphs are erected by the sides of roads, and others pass along railways, or are extended over private property.

The American Government considering that the use of the electric telegraph was a great national question, an Act was passed enabling telegraph companies to construct their apparatus over the lands of public companies or private individuals on the payment of an adequate compensation, which, in case of disagreement between the parties, was to be settled by a jury to be selected for the purpose. Little or no opposition has been shown against

Electric Telegraphs now in Use

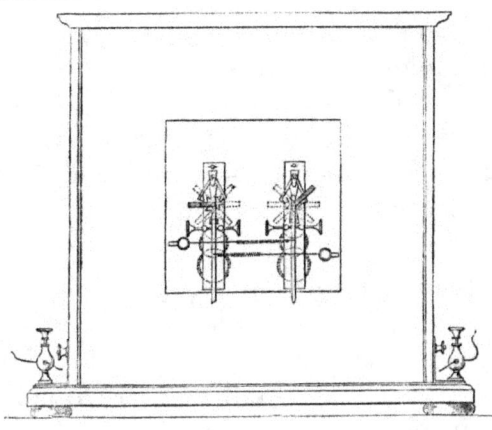

A Front view showing the two indicators

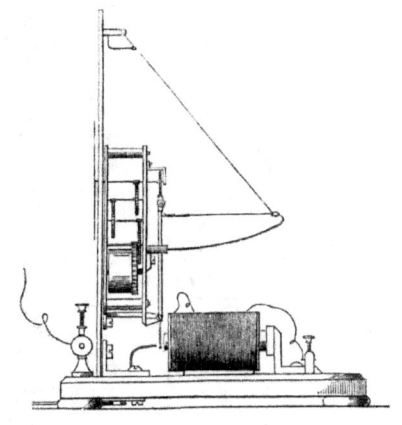

C Side view

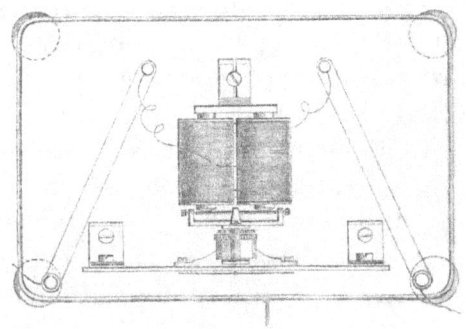

B Top view

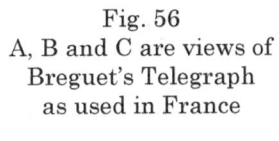

Fig. 56
A, B and C are views of
Breguet's Telegraph
as used in France

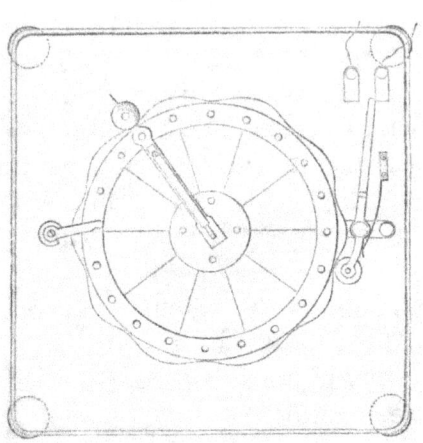

Fig. 57
Breguet's Handle
apparatus for sending
the electric currents

The Electric Telegraph–Its History and Progress

the construction of telegraphs over private property in America. It has been a rule in many cases to allow parties through whose lands the wires pass, and on condition of their repairing the wires when broken, to have their communications sent at a reduced rate.

Telegraphs in France

The telegraphs used in France are for the most part of the class known as revolving pointer telegraphs.

For railway purposes one wire only is used. In this case the pointer, as it revolves, stops successively at the letters on the circular dial which are intended to be denoted.

In the Government and Commercial Telegraphs two wires are used—each wire having its own revolving indicator connected with a separate maintaining power.

Each of these two indicators can be made to rest in any one of 8 different positions in the circle.

It is evident therefore that 8 × 8 or 64 *primary* signals can be transmitted by this form of telegraph.

Figure 56 exhibits this arrangement of instrument by Breguet. The dotted lines on the front view **A** show the various positions in which the two revolving arms can be made to rest. Figure 57 illustrates a Breguet transmitting handle.

Figure 58 shows the alphabetical code used with Breguet's Telegraph, showing the position of the pointers when representing the letters of the alphabet.

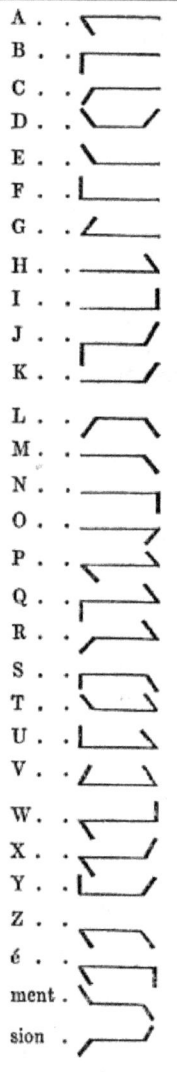

Fig. 58
Alphabetical code used with Breguet's Telegraph, showing the position of the pointers when representing the letters of the alphabet

Electric Telegraphs now in Use

A	.—	O	.———...	CH	—.—..
Æ	.—.—	Œ	.———....	/	—.—...
B	—...	P	.———.	?	...—..
C	—.—.	Q	——.—	1	.———.
D	—..	R	.—.	2	..———.
E	.	S	...	3	...——.
F	..—.	T	—	4	....—
G	——.	U	..—	5	———
H	....	UE	..———	6	......
I	..	V	...—	7	——..
K	—.—	W	.——	8	—....
L	.—..	X	—..—	9	—..—
M	——	Y	—.——	0	—
N	—.	Z	——..		

Fig. 59
Prussian adaptation of Morse code

EXTENT OF TELEGRAPHS IN FRANCE.

In 1850, 620 miles of telegraph were in operation, and 899 miles under construction. The wires are wholly suspended in the air.

A considerable extension of the electric telegraph has taken place since the year 1850, but as no Government return has been published of the number of miles now in use, it is impossible to give the extent of the mileage in operation at the present time.

TELEGRAPHS IN PRUSSIA AND GERMANY.

The wires in Prussia have hitherto been almost universally buried in the earth. Those first used were simply coated with vulcanized gutta percha, and deposited some 18 inches or 2 feet under the surface of the ground.

All telegraphs, however, now under construction, have the gutta percha covered wire encased in a leaden tube. The kind of telegraphic instrument most usually employed is of the class known as the revolving pointer or step-by-step movement telegraph. In this form of telegraph, as the hand revolves it can by mechanism be made to rest, and thereby point out successively the letters on a circular dial. The instruments used are those of Siemens and Fardly. A modification of Morse's telegraph is also employed. Figure 59 is the alphabetical code adopted when Morse's system is employed.

The Electric Telegraph–Its History and Progress

From accounts which have lately arrived in England of the working of the electric telegraphs in Prussia, it appears that the underground system of wires, on the plan first adopted, has proved a failure. It is said that many miles of those subterranean wires are henceforth to be abandoned. In some places wires have already been erected on posts in lieu of them.

In 1850, there were 2,468 miles of wire buried in the earth and in use; 1,210 miles more were then under construction. There are now about 4,000 miles of telegraph at work.

The following statement is extracted from "The Times" of 20th January, 1852:

> The Austrian Government is so disgusted with the snail like way in which the Prussian telegraph does its work, that it is using all its influence to get the line to Strasburg completed. I am credibly informed, that a despatch is frequently 24 hours in getting from one end of the Prussian telegraph line to the other. As the subterranean wires were found to answer so badly when the Emperor was in Italy, they have been entirely abolished in Austria.

It is understood that this refers to the wires *first* laid down, and not to those encased in a leaden tube.

TELEGRAPHS IN RUSSIA AND OTHER COUNTRIES.

Already the ramifications of electro-telegraphs extend from one end of Europe to the other. The lines to connect Petersburg with Moscow, and with the Russian ports on the Black Sea and the Baltic, are in progress. Other wires stretch from the capital of the Czar to Vienna and Berlin, taking Cracow, Warsaw, and Posen on the way. Two lines, by different routes—Olmutz and Brunn,—unite Vienna with Prague, whence an offset leads to Dresden; a third enables the Austrian Government to send messages to Trieste—their outpost on the Adriatic 325 miles distant. A fourth communicates with the metropolis of Bavaria and since the 10th January (1850), the "Gazette d'Augsburg" has published the course of exchange in Munich twenty minutes after it has been declared in Vienna. Calais may send news to the city of the Magyar on the Danube, and ere long intelligence will be flashed without interruption from St. Petersburg to the Pyrenees. Tuscany has 100 miles of telegraph under the direction of Signor Matteucci; and a single wire, traversing the level surface of the Netherlands, unites Rotterdam with Amsterdam. Communities are learning that the electric telegraph is an essential of good government; that police without it is inefficient; that by it the better interests of humanity are promoted. There is talk also of introducing the thought-flasher into that land of wonders—Egypt; to stretch a wire from Cairo to Suez for the service of the overland mail. Who shall say that before the present generation passes away, Downing Street may not be placed in telegraphic *rapport* with Calcutta?

In Austria there are about 3,000 miles of telegraph, one fourth being gutta percha coated wire laid underground.

TELEGRAPHS IN INDIA.

A short time ago Dr. O'Shaughnessy, in India, received instructions from the Indian Government to construct a line of electric telegraph with all possible despatch. At the time at which this order was given, no wire was to be had in India. Dr. O'Shaughnessy was therefore driven of necessity to use such materials as he could obtain in India at the time. He procured a number of rods of iron, $3/8$ ths of an inch thick, and welded them together, in order to form a conductor for the electric current. These rods were joined in lengths of 200 feet before they were conveyed to the line of telegraph. Strong poles were then erected; and these rods of iron, 200 feet in length, being welded together along the line of the electric telegraph, they were erected upon these posts, and thus a continuous metallic conductor was obtained between the distant stations. No sooner was this thick wire erected than monkeys and swarms of large birds alighted on it. No injurious effect, however, was produced upon it by such extraordinary visitors, as the wire was unavoidably composed of so large a diameter.

The telegraph remained perfect for a considerable period, and messages were transmitted with accuracy and despatch. After a time a tremendous storm passed over the district traversed by the wire; a great portion of the posts were prostrated, and the wire thrown down upon the ground. Carts and wagons passed over

The Electric Telegraph–Its History and Progress

the wire without injuring it in the slightest degree. The telegraph was re-erected on stronger posts, and has since performed its work with accuracy and precision. The magnets are placed horizontally, being poised on a vertical axis similar to the mariner's compass. Keys of the simplest construction are used, and highly ingenious arrangements are provided for rendering innocuous the fearful discharges of atmospheric electricity which characterize thunder-storms in the vicinity of the tropics.

The natural currents of electricity are so strong in India that it is impossible to use an electro-magnet of iron in the circuit.

Great credit is due to the energy and perseverance of Dr. O'Shaughnessy, who, without the necessary materials for the construction of an electric telegraph, has contrived, with the scanty materials at his command, to construct one possessing great efficiency combined with great simplicity.

Since the erection of this telegraph, Dr. O'Shaughnessy has arrived in England, under the directions of the East Indian Government, in order to ascertain, from the experience already attained in England and other countries, the best form of telegraph to be used throughout India.

On the return of Dr. O'Shaughnessy to India, upwards of 4,000 miles of telegraph will be constructed with the least possible delay.

One cannot but admire the judicious policy of the East Indian Government in ordering their telegraphic engineer to visit England, for the express object of obtaining the latest improvements in electric telegraphs, before they commence the construction of so great a length of telegraph as that which they contemplate carrying out in their possessions in India.

Whatever may be the cost of this mission, the saving that will thereby be effected will doubtless be enormous, and will tend to the perfection of the system of electric telegraphs throughout these regions of the East.

From the experience that will thus have been gained by the telegraphic engineer of the East Indian Government, one cannot but look forward to the telegraphs in India (where there are no patents) as likely to be the best which have yet been constructed throughout the world, and especially so as these telegraphs will be under the control and direction of Dr. O'Shaughnessy, who for so many years has paid so much attention to the science of electricity as applied to the electric telegraph.

ON THE RESTRICTIONS IMPOSED ON THE USE OF THE ELECTRIC TELEGRAPH

In England, the Government has the power, under various Acts of Parliament, of ordering all messages on Her Majesty's service to take precedence of any other communications whatever.

In cases of civil commotion, or when deemed necessary by the Secretary of State, all telegraphs are required on demand to be placed at the sole disposal and use of the Government. In such cases proper remuneration has to be made to the parties for such use. Only one instance of this kind has occurred, and that was in April, 1849, during the period of the anticipated Chartist riots.

In France, before any despatches are sent they are obliged to be submitted to the Government authorities at the stations, who have full power to refuse or permit their transmission.

All despatches of the Government are transmitted in code or cipher, so that the parties working the telegraph have not the slightest notion of the meaning of the communication.

Cipher or code signals are not allowed to be sent by the public on any pretence whatever.

In Prussia, every communication has to be subjected to similar supervision and control before it is allowed to be sent to the party for whom it is intended. Three distinct systems of signalling by the electric telegraph are used there.

The first system is that in which the ordinary letters of the alphabet are employed, so that everyone can read and understand the communications transmitted. This system is adopted for the transmission of messages for the public, and no other is allowed for such purpose.

In the second system a private code is employed, which code is understood only by the officers in the army.

The third system is by means of a code also. This code differs from the previous one, and is only capable of being translated by a very few of the highest authorities in the Government, such as the Secretary of State and the Commander-in-Chief of the army.

It will be observed therefore that on the Continent the restrictions imposed on the free use, by the public, of the electric telegraph are very great, so much so, indeed, as to render it rather a weapon in the hands of the Governments than a means of promoting social and commercial communications for the community at large.

In America alone, legislative provision is made for the transmission of intelligence for the press out of the due order of reception. All other communications are obliged, under a heavy penalty, to be transmitted in the same order as that in which they are received.

In England, penalties are also imposed by Acts of Parliament for the transmission of any communications otherwise than in the order of their reception, but no exception to this rule is made with regard to communications intended for the press.

TELEGRAPHS IN THE BANK OF ENGLAND.

The Court of Directors of the Bank of England have lately caused to be erected throughout their buildings a system of electric telegraphs, to communicate between the different departments of that establishment.

These telegraphs have been erected partly under the patents of Mr. Dering, which had been previously bought by the old Electric Telegraph Company, Mr. Dering himself either being then or about to become a Director of that Company.

A description of the patent of Mr. Dering would have been given in an earlier portion of this treatise, if these telegraphs had been in use when such former part was written.

The principal features in the plans of telegraphs patented by Mr. Dering may be thus summed up:

1. Instead of the moveable magnets working on axes, magnets fixed to elastic supports are employed.

Electric Telegraphs now in Use

2. The magnet and coil are so arranged that the electricity passing through the coil acts on the magnet by attraction or repulsion; the coil resembling that of an electro-magnet, but without the iron core.
3. A peculiar method of sounding alarums is also given.
4. A means of cutting off from the circuit any particular telegraphic instrument is described.
5. Several methods of preventing the injurious effect of atmospheric electricity, collected by the wires of the telegraph during thunderstorms, are given.
6. In insulating the line-wires an additional bell insulator is employed, this second bell being inverted and placed inside the usual one.

The Specification of this patent is published at length in the "Repertory of Patent Inventions." Mr. Dering's telegraphs are now also partially used on the Great Northern Railway.

ELECTRIC TELEGRAPHS IN AMERICA IN CONNECTION WITH THE FIRE ESTABLISHMENTS

In some of the principal towns in America a system of electric telegraphs has been constructed, by means of which notice of the breaking out of a fire is instantly telegraphed to all the fire-engine stations.

A town is divided into certain districts. In each district one or more electric telegraphs are placed, communicating with a central station. From this central station wires diverge to all the principal establishments where fire-engines are kept.

As soon as a fire breaks out, information is sent to the central police station. From this central station communications are instantly despatched by telegraph to all the fire-engine stations. The engines start immediately to the scene of the fire. Assistance is thus despatched forthwith, and no delay arises from the uncertainty as to the exact locality of the fire.

It is evident that if such a system were adopted in London a large amount of property might be annually saved, and the loss of many lives prevented.

It is hoped, therefore, that the inhabitants of this great metropolis will not suffer themselves to remain long behind their enterprising brothers on the other side the Atlantic, in a matter so clearly affecting not only their property, but even their own lives.

11

SUBMARINE TELEGRAPHS

In 1849, an English Company obtained a charter from the French Government, which granted to them the exclusive right for 10 years of sending electro-telegraphic intelligence between England and certain defined points on the French coast. This right was granted on condition that certain requirements were complied with, and the work carried out within a given period.

The first line laid down consisted of one copper wire simply covered with gutta percha. This wire was laid across the Channel in August, 1850. The covering of the gutta percha was $1/4$ of an inch thick.

The wire remained perfect, however, only a few hours, as the action of the sea rolling it about on the sharp rocks at once destroyed the covering and rendered the wire useless.

In September, 1851, another line of telegraph was laid across the English Channel. This consisted of four copper wires, each encased in gutta percha, and then enclosed in a rope of galvanized iron. The length of rope made was 24 miles. It weighed when finished 180 tons. The plan adopted in the manufacture of this telegraph cable was as follows:

A copper wire (No. 16 wire gauge) was first carefully covered with gutta percha; upon this coating of gutta percha a second covering was laid; the copper wire was thus thoroughly well insulated. Four of these insulated wires were then bound together with spun-yarn and hemp, saturated with tar.

This bundle of insulated wires with its hempen covering was then surrounded by ten galvanized iron wires, each wire being $5/16$ ths of an inch in diameter. The insulated wires thus formed the core of a large wire-rope; the whole process and the principle employed being both exactly the same as those patented by the

Fig. 60
Cable laid in the English Channel between England and France. September 1861

author in 1850.

As this telegraphic wire-rope came from the machine, it was formed into a large coil 30 feet in diameter. Each of the external and internal wires were in one unbroken length. The several smaller lengths of the external wire, as manufactured, were welded together, and the inner ones soldered. The making of the rope occupied twenty days.

Figure 60 shows a portion of the rope as finished.

The machine which laid the iron wires around the insulated ones, made, when working freely, about 18 revolutions per minute, and completed about 11 inches of the cable in that time.

This huge wire-rope was then shipped on board the "Blazer," an old war steamer, which the Admiralty placed at the disposal of the Company. The machinery of the steamer, together with the funnel, had all been previously removed in order to obtain sufficient space in the hull for the coil of the cable. The "Blazer" was then towed from London to Dover.

On the 25th of September, 1851, the work of paying out the cable commenced. Steam-tugs were placed by the Admiralty at the command of the Company. The "Blazer," with her cargo, was then towed from Dover to the South Foreland, and one end of the rope conveyed on to the English shore. After this the vessel was towed in the direction of Cape Griz Nez.

During the process of paying out the rope, it appears that many kinks or bends occurred, and the covering was every now and then torn off the insulated wires as the rope went through the opening made for it in the vessel. So great was the damage done at one time, that it was thought that the inner telegraphic wires were greatly injured. On the testing of them, however, the insulation was found to be perfect. It is hoped that time may not reveal the fact of the insulating covering having in any way been seriously injured.

The distance between the extreme points on the two coasts between which the cable was to extend, was 20 miles. An extra length of 4 miles of cable was made to allow for undulations and sinuosities. In consequence of the manner in which the cable was put on board the steamer and afterwards payed out, and the sinuosities of the course traversed by the vessel (which at one time broke away from the steam tugs), the extra length of 4 miles of rope, as allowed in its length, was found to be too little. The end of the 24 miles of rope would not reach its destination by about half-a-mile.

After temporarily connecting the wires in the cable to some spare wire simply covered with gutta percha, and thereby passing a few complimentary messages from coast to coast, operations were suspended until more cable could be manufactured. Another mile of the same kind of cable was made, spliced to the end of the old one, and then laid down in the sea.

On the 18th of October the communication was found to be perfect.

The cost of the cable is said to have been £20,000, and the whole expenses of the Company no less than £75,000.

Arrangements are being made for trying, through the instrumentality of the submarine telegraph, some remarkably curious astronomical experiments, and it is considered that facilities for sidereal observation on all parts of the Continent will be greatly increased by means of it. The South-Eastern Railway Company, have, it is said, with a view to the promotion of this object, consented to carry a wire or wires from their telegraph to the Observatory at Greenwich, so as to connect it with the submarine wires, and thus with the Observatory at Paris, so that simultaneous observations may be made between the Astronomer Royal here and Professor Arago in Paris. The transit of a star over the meridian of London and Paris can thus be made and notified in an instant, and with it the exact time of its transition. The longitude of both places, and of different places on the Continent, can also be easily obtained, and the most accurate records of comparative astronomy be recorded and preserved.

The working of these submarine wires has hitherto proved very satisfactory—and, it is stated, highly remunerative to the proprietors.

A second rope is now under construction for laying down between England and France, as a duplicate to the first one.

Another telegraphic rope is about to be laid down between England and Ostend. An exclusive permission to do so has already been accorded to the principal proprietors in the submarine telegraph from England to France. A third rope is also to be submerged between England and Holland.

A submarine rope with one insulated wire has lately been laid down between Holyhead and Dublin, and other similar ropes will shortly be submerged between Port Patrick and Donaghadee.

The submarine wire-rope lately laid across the Irish Channel between Dublin and Holyhead was thus constructed—a copper wire, No. 16, was covered with two coatings of gutta percha. In that part of the rope which lies in deep water, the gutta percha covering is merely surrounded by twelve No. 16 galvanized iron wires, forming a thin rope about the size of one's little finger. Near the shore the gutta- percha was protected by a covering of six very thick galvanized iron wires. The weight of the rope was about 80 tons, and the length 80 miles. This rope has at present proved an entire failure, and many miles of it have already been taken up.

TELEGRAPHING WITHOUT INSULATION.

Various experiments have been tried in England, America, and India, with a view to ascertain whether it was possible to send telegraphic communications with naked wires, or even without any wires at all.

In England the author and his brother have tried many experiments on this subject. Naked wires have been sunk in canals, for the purpose of ascertaining the mathematical law which governs the loss of power when no insulation was used. Communications were made with ease over a distance of about a quarter of a mile.

The result, however, of these experiments has been to prove that telegraphic communications could not be sent to any considerable distance without the employment of an insulating medium.

In India Dr. O'Shaughnessy has laid uninsulated wires across a river which is more than a mile broad, for the purpose of transmitting telegraphic communications, and he has found that to transmit a current along an uninsulated wire of that length, and to obtain at the distant end an action sufficient to work his telegraphic instruments, no less than 250 galvanic cells were required, and that even then the signals were scarcely visible.

Professor Morse, in America, has tried various experiments in sending currents across rivers without any intervening wires.

All these experiments, however, have led to no practical result, except that they have proved that where it is necessary to transmit currents of electricity between two stations which are far distant from each other, insulation of a metallic medium is absolutely required. Beyond the proof of this fact, all these costly experiments have proved futile.

The experiments themselves have, however, been of the highest value to the science of telegraphing by electricity, and many laws have been deduced therefrom.

12
MESSAGE CHARGES AND THE REGULATION OF TIME

Charges for the Use if the Electric Telegraph

When the old Electric Telegraph Company first opened their lines for the transmission of public messages in England, the charges for twenty words were calculated at the rate of 1d. per mile for the first 50 miles, ½d. per mile for the next 50 miles, and ¼d. per mile for every mile beyond the first 100 miles.

- On the 11th of March, 1850, the charges were reduced; 10s. being made the maximum charge for *any* distance.
- On the 20th of March, 1851, a further reduction was made, and no message of twenty words was to exceed 8s. 6d.
- On the 17th of November, 1851, the tariff was still further reduced; the charge being 2s. 6d. for twenty words if transmitted 100 miles or less, and 5s. if more than 100 miles.
- Early in 1852 a further reduction was made, the charge for a message between Manchester and Liverpool being for twenty words 1s. instead of 2s. 6d.

The telegraphs on the South-Eastern Railway do not belong to the old Electric Telegraph Company, but to the Railway Company. The charges in the district passed through by this railway were very high in the first instance, but a great reduction was made on the 17th of November, 1851, when the charge for a message between any two stations on their line was reduced to 5s. for twenty words.

Time and experience alone can decide whether the charges now made in England are such as to produce a maximum of convenience to the public, together with a maximum of profit to the telegraph companies. There are some who assert that the above charges are very high, and that neither the public reap the benefits which they ought to derive from so great an invention, or the proprietors of the telegraph so great a profit as they would do if the charges were materially reduced. There are others, again, who assert that the present charges are as low as they ought to be to secure to the telegraph company a fair return on the capital invested by them in the construction of their lines, and at the same time to give considerable facilities and convenience to the public. Time, experience, and competition will ere long afford a perfect solution to this problem. Much, it is true, will ever depend on the cheap construction of the electric telegraph itself, although dependence must not be placed wholly on this point.

Accuracy, precision, and despatch are of far greater importance in the transmission of telegraphic intelligence than mere cheapness; but there is no reason why all these essential properties should not be combined with low charges.

Another point which will in a great measure regulate the charge for the transmission of telegraphic messages in ordinary matters is the increasing facilities and despatch which, by means of improvements in railways, can and will be afforded to the transmission of intelligence by the post.

Immense improvements have of late years been made in the despatch of letters by post. This department is under the control of Government, and increased facilities for the transmission of letters are taking place almost daily. The charge for telegraphic intelligence must therefore in some measure be regulated by the increased facilities which may gradually be afforded by the Post Office. That charge which may be the best for all parties to-day may require alteration on the morrow; no definite general rule, therefore, can be laid down.

The charge for the use of the telegraph must, as in the case of the fares by railway, steamboat, or omnibus, be regulated by the peculiar circumstances existing at the time.

No one, however, can doubt, that as in the case of the late reduction in the charge of postage for letters the number of letters

Message Charges and the Regulation of Time

immensely increased, so, in the reduction of charges for the use of the telegraph, the number of messages would be greatly multiplied also. On pages 164/5 are shown the present tariff of charges for the transmission of messages by the submarine telegraph.

On page 166 are shown the charges for the use of the telegraph in Belgium in 1850.

TELEGRAPH CHARGES IN FRANCE AND PRUSSIA.

The charges for the transmission of telegraphic intelligence in France and Prussia were, in 1850, about the same as those in England at that time.

The following Table is the tariff of charges in France:

			miles.	f.	c.	s.	d.
Paris to	Amiens	about	72	4	80	4	0
"	Arras		100	5	64	4	8 3/5
"	Valenciennes		125	6	36	5	3 3/5
"	Lille		130	6	36	5	3 3/5
"	Calais		163	7	56	6	3 3/5
"	Dunkirk		160	7	32	6	1 1/5
"	Orleans		67	4	56	3	11 3/5
"	Tours		139	5	88	4	10 4/5
"	Angers		190	7	60	6	4
"	Bourges		127	5	88	4	10 4/5
"	Nevers		154	6	72	5	7 1/5
"	Chateauroux		147	6	24	5	2 2/5
"	Chalons sur Marne		95	5	10	4	3
"	Rouen		75	4	68	3	10 4/5
"	Havre		112	5	76	4	9 3/5

In America the charges have always been very low, and the telegraph in consequence is most extensively used. In 1850 the charges varied from one-fourth to one-sixth of those made in England.

SUBMARINE TELEGRAPH.

Rates for Transmission of Messages.

FROM LONDON	NUMBER OF WORDS.								
	20	21 to 30	31 to 40	41 to 50	51 to 60	61 to 70	71 to 80	81 to 90	91 to 100
	£ s. d.	£ s. d.	£ s. d.	£ s. d.	£ s. d.	£ s. d.	£ s. d.	£ s. d.	£ s. d.
To Amiens	0 17 6	1 2 8	1 5 0	1 8 0	1 11 6	1 14 6	1 18 0	2 1 0	2 4 6
" Angers									
" Arras	0 17 0	1 1 0	1 13 0	1 17 0	1 11 0	1 14 0	2 1 0	2 15 0	2 3 0
" Blois									
" Bourges	1 15 6	1 7 0	1 10 0	1 14 0	1 19 0	2 4 0	2 7 0	2 11 0	2 15 0
" CALAIS	0 12 6	0 16 0	1 11 18	1 15 0	1 19 2	2 2 4	2 5 6	2 8 8	2 11 10
" Chalons, S.M.									
" Chalons, S.S.	1 2 6	1 5 3	1 9 6	1 13 6	1 17 0	2 0 6	2 5 6	2 9 6	2 13 0
" Chateauroux	1 3 6	1 8 7	1 12 0	1 17 6	1 10 2	2 5 2	2 9 6	2 15 3	2 19 0
" Dunkirk	0 16 0	0 18 6	1 3 2	1 6 6	1 9 6	1 12 6	1 14 6	1 17 6	2 0 6
" Dijon	1 2 0	1 8 0	1 12 9	1 16 0	1 9 2	2 5 0	2 9 6	2 14 0	2 18 6
" Havre									
" Lisle	0 16 0	1 0 5	1 3 8	1 17 0	1 9 6	1 12 5	1 14 4	1 17 8	2 0 0
" Nevers	1 2 6	1 2 0	1 12 0	1 17 6	1 9 6	2 5 0	2 4 6	2 9 0	2 19 6
" Orleans									
" PARIS	0 19 0	1 3 6	1 7 0	1 10 6	1 14 6	1 17 6	1 1 6	2 4 8	2 8 6
" Rouen	0 14 6	1 0 6	1 9 0	1 13 6	1 17 0	2 0 2	2 4 6	2 9 0	2 12 6
" Tonnerre									
" Tours	1 2 6	1 7 8	1 11 0	1 15 0	1 14 0	1 17 0	2 7 6	2 11 0	2 15 0
" Valenciennes	0 17 6	0 18 6	1 4 0	1 7 0	1 10 2	1 13 6	1 16 0	1 19 6	2 2 0
" Poictiers									

,, Nantes	1	3	6	1	9	6	1	14	0	1	19	0	2	3	0	2	7	6	2	13	0	2	17	6	3	3	0
,, QUIEVRAIN (Belgian Frontier)	0	17	0	1	1	0	1	4	6	1	7	6	1	10	6	1	13	6	1	16	6	2	0	0	2	3	6
,, BRUSSELS, ANTWERP, Ghent, Malines, and Bruges	1	1	0	1	9	6	1	12	6	1	15	6	2	2	6	2	5	6	2	8	6	2	11	6	2	14	6
,, Mons	0	19	6		5	6	1	9	0	1	12	0	1	17	6	2	1	0	2	4	0	2	7	6	2	10	6
,, Verviers, OSTEND	1	3	0	1	13	0	1	16	6	1	19	6	2	8	6	2	11	6	2	14	6	2	17	6	3	0	6
,, Aix-la-Chapelle	1	5	0	1	17	0	2	0	6	2	3	6	2	14	6	2	17	6	3	0	6	3	3	6	3	6	6
,, COLOGNE, Deutz, Elberfeld, Dusseldorf, Duisburg	1	7	0	2	1	6	2	4	6	2	7	6	3	2	6	3	4	6	3	6	6	3	9	6	3	13	0
,, Munster, Ham	1	9	0	2	5	6	2	8	6	2	11	6	3	6	6	3	9	6	3	12	6	3	15	6	3	18	6
,, HANOVER, Brunswick, Minden	1	11	0	2	9	6	2	12	6	2	15	6	3	12	6	3	15	6	3	18	6	4	1	6	4	4	6
,, BERLIN, Cothen, Dessau, Halle, Jüterburg, Leipsic, Magdeburg, Oschersleben, Potsdam . .	1	12	6	2	13	6	2	16	6	2	19	6	3	18	6	4	1	6	4	4	6	4	7	6	4	10	6
,, Altenburg, Bodenbach, Buchen, Cassel, DRESDEN, Eisenach, Erfurt, FRANKFORT-SUR-ODER, Gotha, Hagenau, Hof, Kreutz, Plauen, STETTIN, Swinnemunde, Weimar, Werden, Wittenburg	1	15	0	2	16	6	2	19	6	3	2	6	4	4	6	4	7	6	4	10	6	4	13	6	4	16	6
,, Aschaffenburg, Augsburg, Bamburg, Bromburg, Breslau, Brunn, Cosel, FRANKFORT-ON-THE-MAINE, HAMBURG, Hanau, Geissen, Liegnitz, Marburg, Nuremburg, Oderburg, Olmutz, Oppeln, POSEN, PRAGUE, Ratibor, Trubau, Wurzburg	1	17	0	3	1	6	3	4	6	3	7	6	4	10	6	4	13	6	4	16	6	4	19	6	5	2	6
,, Bielitz, CRACOW, Friedrichshafen, Heilborn, Linz, MUNICH (by Hof), Neuhausel, PRESBURG, STUTTGARD, Salzburg, Troppau, Ulm, VIENNA	1	19	0	3	5	0	3	8	0	3	11	0	4	16	6	4	19	6	5	2	6	5	5	6	5	8	6
,, Agram, Bregentz, Brixen, Botzen, Cilly, Feldkirk, Gratz, Inspruch, Kufstein, Laibach, PESTH, Pirano, Roveredo, Trento, TRIESTE, VERONA	2	1	0	3	9	0	3	12	0	3	15	0	5	2	6	5	5	6	5	8	6	5	11	6	5	14	6
,, Brescia, Goritz, MANTUA, MILAN, PADUA, Peschiera, Trevisa, Udine, VENICE, Vicenza .	2	3	0	3	13	0	3	16	0	3	19	0	5	8	6	5	11	6	5	14	6	5	17	6	6	0	6

N.B. The above rates are exclusive of the usual charge for Porterage for the delivery of the Messages.

1st February, 1852. LEWIS C. HERTSLET, *Secretary*.

TELEGRAPH CHARGES IN BELGIUM.

	French Measures.				In *English* Measures and English Money.			
		Communications.				Communications.		
Distances.	1 and not exceeding 20 words.	21 and not exceeding 50 words.	51 and not exceeding 100 words.		Distances.	1 and not exceeding 20 words.	21 and not exceeding 50 words.	51 and not exceeding 100 words.
	f. c.	f. c.	f. c.			£ s. d.	£ s. d.	£ s. d.
For 1 and not exceeding 75 kilometres	2 50	5 00	7 50		For ⅗ths of a mile & not exceeding 46⅞ miles	0 2 1	0 4 2	0 6 3
For 76 and not exceeding 200 kilometres	5 00	10 00	15 00		For 47½ miles and not exceeding 125 miles	0 4 2	0 8 4	0 12 6
Above 200 kilometres	7 50	15 00	22 50		Above 125 miles	0 6 3	0 12 6	0 18 9

Message Charges and the Regulation of Time

The following Table shows a few of the charges made in America about that time:

			miles			s.	d.
From	New York to Boston	about	240	30 cents, or about		1	3
	Philadelphia to Harrisburg	"	107	20 "	"	0	10
	New York to Philadelphia	"	90	25 "	"	1	0
	Philadelphia to Pittsburg	"	321	40 "	"	1	8
	New York to Baltimore	"	190	50 "	"	2	0
	New York to Albany	"	196	37 "	"	1	6
	New York to Washington	"	230	50 "	"	2	0
	Washington to New Orleans	"	1716	2 dollars,	"	8	0
	Washington to Baltimore	"	40	20 cents	"	0	10
	Philadelphia to Wilmington	"	30	10 "	"	0	5
	Wilmington to Baltimore	"	86	20 "	"	0	10
	Philadelphia to Baltimore	"	99	25 "	"	1	0

Longer messages are sent at proportionally low rates, with a further reduction on very long communications. In contrast with this, the charges in England were at the same period:

						s.	d.
From London to	Birmingham, which is about	112	miles	6	6
"	Cheltenham,	"	100	"	7	6
"	Glasgow	"	420	"	10	0
"	Hull	"	200	"	9	6
"	Liverpool	"	210	"	8	6
"	Newcastle	"	300	"	10	0
"	Southampton	"	80	"	5	6
"	York	"	200	"	9	0

It should be observed that these English rates are for a message of twenty words, whilst the American rates are for a message of ten words: but then the English Company count the address and signature, &c. as forming part of the twenty words; whilst the American rates relate to a message of ten words, clear of the address, date, signature, &c., for which nothing is charged.

REGULATION OF TIME BY THE TELEGRAPH

For some time past arrangements have been pending between the Electric Telegraph Company, the Astronomer Royal, and the South-Eastern Railway Company, for the establishment and transmission, throughout London and the provinces, of mean Greenwich or uniform time. For this purpose a system of ingenious apparatus is being constructed upon the dome of the Telegraph Company's West End station, No. 448, Strand, opposite Hungerford Market. From the summit of this dome an uninterrupted view of London and the river was obtainable, the total height of the apparatus being about 110 feet above the level of the Thames.

The apparatus consists of a long quadrangular shaft or pillar of wood, about 38 feet high. The first section of this hollow shaft is fixed into the floor of the room underneath the dome, and thence carried through and joined on to the second section and the third, the latter of which is then passed through the centre of a large globe or ball, which is intended by means of sympathetic electrical action to fall every day simultaneously with the well-known ball on the top of the Greenwich Observatory, between which and the Strand the electric wires have been completed for the purpose, so as to indicate to all London and the vessels below London bridge exact Greenwich time. The ball is nearly 6 feet high and 16 feet in circumference, and is formed of zinc. The apparatus is so constructed, in connection with the telegraphic wires between London and Greenwich, that when the ball at Greenwich falls, an instantaneous shock of electricity will be communicated along them; and this, acting on an electrical trigger connected with the ball in the Strand, will cause it to fall simultaneously with the one at Greenwich. The cost of carrying out this novel chronometrical machine is estimated at £1,000.

ELECTRIC CLOCKS.

Various plans have been proposed and patented for working Clocks or chronometers by the power of electricity instead of that of gravity. In some cases no weight is employed, but the power of electricity is the direct means of causing a pendulum to continue its vibrations for any definite period of time, and thus to force forward the wheels of clockwork mechanism, and with them the hands of the clock.

Among the inventors of this system of electric clocks may be mentioned with distinction the names of Mr. Alexander Bain and Mr. Shepherd.

Mr. Bain and others have also caused the motion of clockwork mechanism, whether the same be actuated primarily by electricity or gravity, to send currents of electricity along a conductor, and thus to work any number of electric clocks placed in the circuit, and thereby to make each one of those clocks to keep exactly the same time as that of the primary or standard one.

This principle of causing any number of clocks to go isochronously is of the highest importance, and especially so in Observatories, where observations are being made at the same instant of time on the same celestial object, by different observers at different instruments.

Electric Clocks at the Observatory at Greenwich.

At the present time several clocks on this latter principle are under construction at the Observatory at Greenwich.

Hitherto at this Observatory each attendant or observer has had his own chronometer; and as no two chronometers working by the power of gravity or a spring can be made to go exactly isochronously for weeks together, each observer in the different departments, after having denoted the time of an observation, has had to correct such denoted time, in order to make it correspond with the time of the standard chronometer of the Observatory.

As soon, however, as every clock in the building is made by the means of the electric current to go isochronously with the standard chronometer of the Observatory, no such correction will be required, nor indeed any correction at all, unless the standard chronometer happens at the time to be either before or behind the *true* time. This arrangement of chronometers, actuated by means of electric currents, will, when carried into practice, immensely reduce the labours of each observer, and give far more accurate results.

An apparatus of another kind is now being constructed at the Observatory, by means of which the exact instant of time at which a heavenly body crosses the wires of the telescope may be denoted with absolute certainty.

The apparatus consists of a cylinder, in circumference about 3 feet, and over which a covering of paper is to be placed. This cylinder will be made to revolve uniformly by means of a clock movement, governed by a centrifugal pendulum. The paper will travel a distance of about one-third of an inch in a second of time, but of course it can be made to travel a greater or less distance if desired. A pricker or marker, to be actuated by means of electricity, will be placed immediately over this paper. Wires will extend from this marker, and terminate in a key by the side of the observer at the telescope. On pressing down this key a mark will be made on the paper on the revolving cylinder, in the same way as marks are made by Morse's telegraph.

At the instant that the observer notices that a heavenly body is crossing one of the cross-hairs in his telescope he will touch the key, and thus cause a mark to be made on the moving paper. Each second of time will, by clock-work mechanism, be also marked by dots upon the same piece of paper, so that by measuring the distance of the dot made by the observer from the nearest second dot made by the clock-work, the exact instant of time at which that observer's mark was made will be ascertained.

By this means the observer will be enabled to record the exact period at which such heavenly body crosses the respective cross-hairs of his telescope, and that with a degree of nicety which can be measured to the fractional part of a second of time.

Without such apparatus it would be almost impossible to denote the time of an observation to a degree much less than that of a whole second of time, but by its means it will be perfectly easy to ascertain the exact instant, even to the $1/100$ th part of a second.

It is obvious, then, that by the aid of electricity all astronomical observations will henceforth be conducted with a degree of accuracy which no one a few years ago would ever have imagined to have been practically possible.

To such a degree of nicety has this means of measuring

Message Charges and the Regulation of Time

extremely small periods of time been carried in other departments of science, that the author himself has been enabled to record the time at which occurrences have taken place to the $1/1000$ th part of a second of time. Such has been the late rapid advancement in the science of Electricity.

If anyone had dared but a few years ago to have asserted that he could have recorded the period of an occurrence to $1/1000$ th part of a second of time, he would have been looked upon as an idle dreamer, or one devoid of common sense. The fact is, however, patent to the world. To the uninitiated, the measuring of such extremely small periods of time may even now appear chimerical, but to the man of science such has long become an acknowledged fact.

In the United States the principle of telegraphing time by electricity has long been adopted. At Boston, U. S., true time is received every day from the Cambridge Observatory, four miles distant, for the service of the shipping in the harbour.

There is no reason why, in this great commercial Country, Greenwich time should not be sent at least once every day to every shipping port in the United Kingdom. The advantages which such an arrangement would afford are beyond calculation, and it well becomes the Government of this maritime nation to see that such an arrangement is speedily and effectually carried out.

13

UTILITY OF THE ELECTRIC TELEGRAPH

No person unacquainted with the electric telegraph can form any idea of the enormous amount of business which is daily transacted by means of this invention. Messages of every character are constantly being transmitted, both by night as well as by day.

As regards the amount of telegraphic correspondence carried on by a railway, with respect to the working of the line alone, some idea may be formed by the knowledge of the fact that at a single station, on the South-Eastern Railway, upwards of 20,000 messages on the service of the Railway are annually recorded in the books of that Company.

The security which the electric telegraph affords to railway travelling is not the least of its merits; accident and obstruction can at once he made known, and the remedy provided for.

The following is one of the many services which the electric telegraph has rendered, in averting impending danger on a railway.

"On New Year's Day, 1850, a catastrophe, which it is fearful to contemplate, was averted by the aid of the telegraph. A collision had occurred to an empty train at Gravesend; and the driver having leaped from his engine, the latter started alone at full speed to London. Notice was immediately given by telegraph to London and other stations; and while the line was kept clear, an engine and other arrangements were prepared as a buttress to receive the runaway. The superintendent of the railway also started down the line on an engine; and on passing the runaway he reversed his engine, and had it transferred at the next crossing to the up-line, so as to be in the rear of the fugitive. He then started in chase, and on overtaking the other he ran into it at speed, and the driver of his engine took possession of the fugitive, and all danger was at an end. Twelve stations were passed in safety; it went by Woolwich at fifteen miles an hour, and was within a couple of miles of London before it was arrested. Had its approach been unknown, the mere money-value of the damage it would have caused might have equalled the cost of the whole line of telegraph."

The promptitude with which detection has followed fraud by the agency of the telegraph is sometimes rather amusing. Mr. Smee relates an instance:

"One Friday night, at ten o'clock, the chief cashier of the Bank received a notice from Liverpool, by electric telegraph, to stop certain notes. The next morning the descriptions were placed upon a card and given to the proper officer, to watch that no person exchanged them for gold. Within ten minutes they were presented at the counter by an apparent foreigner, who pretended not to speak a word of English. A clerk in the office, who spoke German, interrogated him, when he declared that he had received them on the Exchange at Antwerp six weeks before. Upon reference to the books, however, it appeared that the notes had only been issued from the Bank about fourteen days, and therefore he was at once detected as the utterer of a falsehood. The terrible Forrester was sent for, who forthwith locked him up, and the notes were detained. A letter was at once written to Liverpool, and the real owner of the notes came up to town on Monday morning. He stated that he was about to sail for America, and that whilst at an hotel he had exhibited the notes. The person in custody advised him to stow the valuables in his portmanteau, as Liverpool was a very dangerous place for a man to walk about with so much money in his pocket. The owner of the property had no sooner left the house than his adviser broke open the portmanteau and stole the property. The thief was taken to the Mansion House, and could not make any defence. By a little after ten the next morning, such was the speed, not only was a true bill found, but the trial by petty jury was concluded, and the thief sentenced to expiate his offence by ten years' exile from his native country."

Again:

"When the 'Hibernia' steamer arrived at Boston in January, 1847, with news of the scarcity in Great Britain, Ireland, and other parts of Europe, and with heavy orders for agricultural

Utility of the Electric Telegraph

produce, the farmers in the interior of the State of New York, informed of the facts by magnetic telegraph, were thronging the streets of Albany with innumerable team-loads of grain almost as quickly after the arrival of the steamer at Boston as the news of that arrival could ordinarily have reached them."

Apart from business and politics, the Americans have made the electric telegraph subservient to other uses: medical practitioners in distant towns have been consulted, and their prescriptions transmitted along the wire; and a short time since a gallant gentleman in Boston married a lady in New York by telegraph—a process which may supersede the necessity for elopement, provided the law hold the ceremony valid.

A favourable idea of the immediate practical utility of the telegraph may be gathered from the remarks of a correspondent in the United States, who writes thus:

> "The telegraph is used in this country by all classes, except the very poorest—the same as the mail. A man leaves his family for a week or a month; he telegraphs them of his health and whereabouts from time to time. If returning home, on reaching Albany or Philadelphia, he sends word the hour that he will arrive. In the towns about New York the most ordinary messages are sent in this way; a joke, an invitation to a party, an inquiry about health, &c. In our business we use it continually. The other day two different men from Montreal wanted credit, and had no references; we said, 'Very well; look out the goods, and we will see about it.' Meanwhile we asked our friends in Montreal, 'Are Pump and Proser good for one hundred dollars each'. The answer was immediately returned, and we acted accordingly; probably much to our customers' surprise. The charge was a dollar for each message, distance about 500 miles, but much further by telegraph, as it has to go around to avoid the water."

In that excellent publication, viz. "Chambers' Papers for the People" from which the above quotations have been made, appear the following verses on the electric telegraph. They are inserted here merely for the purpose of showing the varied uses to which the electric telegraph is even now employed.

The Electric Telegraph–Its History and Progress

Hark! the warning needles click,
Hither—thither—clear and quick.
Swinging lightly to and fro,
Tidings from afar they show,
While the patient watcher reads
As the rapid movement leads.
He who guides their speaking play
Stands a thousand miles away.
Sing who will of Orphean lyre,
Ours the wonder-working wire!
Eloquent, though all unheard,
Swiftly speeds the secret word,
Light or dark or foul or fair,
Still a message prompt to bear:
None can read it on the way,
None its unseen transit stay.
Now it comes in sentence brief,
Now it tells of loss and grief,
Now of sorrow, now of mirth,
Now a wedding, now a birth,
Now of cunning, now of crime,
Now of trade in wane or prime,
Now of safe or sunken ships,
Now the murderer outstrips,
Now it warns of failing breath,
Strikes or stays the stroke of death.
Sing who will of Orphean lyre,
Ours the wonder-working wire!
Now what stirring news it brings,
Plots of emperors and kings;
Or of people grown to strength,
Rising from their knees at length:
These to win a state—or school;
Those for flight or stronger rule.
All that nations dare or feel,
All that serves the commonweal,
All that tells of government,

On the wondrous impulse sent,
Marks how bold Invention's flight
Makes the widest realms unite.
It can fetters break or bind,
Foster or betray the mind,
Urge to war, incite to peace,
Toil impel, or bid it cease.
Sing who will of Orphean lyre,
Ours the wonder-working wire!
Speak the word, and think the thought,
Quick 'tis as with lightning caught,
Over—under—lands or seas,
 To the far antipodes.
Now o'er cities throng'd with men,
Forest now or lonely glen;
Now where busy Commerce broods,
Now in wildest solitudes;
Now where Christian temples stand,
Now afar in Pagan land.
Here again as soon as gone,
Making all the earth as one.
Moscow speaks at twelve o'clock,
London reads ere noon the shock;
Seems it not a feat sublime,—
Intellect hath conquer'd Time!
Sing who will of Orphean lyre,
Ours the wonder-working wire!

Concluding Remarks.

In penning the foregoing pages, the author has endeavoured to free his mind as much as possible from any bias that it might have in favour of any particular system or systems of electric telegraph.

As the parent naturally likes his own offspring better than the child of another, so each inventor naturally considers his own inventions superior to those of his neighbour; and should any remarks that have been made be thought to be in any way uncalled for, or harsh, or severe, the author will be truly and sincerely sorry. Difficult though the task in practice may be, his sincere aim and object have ever been to do justice to all.

Many persons may think that some inventions ought to have been given more *in extenso*, while others ought to have been curtailed. To please all would clearly have been impossible, and the author trusts that those who are inclined to censure will remember the fable of the old man, the boy, and the ass.

It must not be forgotten that the electric telegraph is now only in its infancy. Much has yet to be accomplished, and wonderful discoveries still await the patient and laborious student.

Who could have thought that the accidental contraction of the muscles of a frog would ever have paved the way to such brilliant results as have already appeared!

The falling apple led to the discovery of the power of gravity, and what may not a further search into the hidden mystery of electricity ultimately bring to light!

Electricity and steam have now become the great civilizers of mankind. Time and space are all but annihilated. Years are converted into days, days into seconds, and miles have become mere fractions of an inch.

No fairy dream could ever surpass the wonders of the present age. The ray of light is caused—*itself* to paint the verdant landscape, or trace the features of our dearest friends. The power of heat is made to carry us whithersoever we will—it transports alike the luscious fruits of the tropics to the colder regions of the earth, or carries the cooling ice to assuage the parching thirst of sunny climes, and last, but not least, that tiny thread of wire which dangles in the air conveys a silent current far away, and thus transacts the business of a mighty trade and commerce, and carries our very thoughts to dear and distant friends. All this is the work of the last half century. Who can tell what another century may bring forth!

Search further, ye patient labourers in the field of electricity, for many brilliant discoveries await you still. Examine well the laws that govern this subtle power, and soon will you meet your reward. The field is rich in the extreme. Careful study, and direct experiments, are alone required to add fresh laurels to the many that already adorn the brows of the Philosophers of the present age. It is time to conclude; in doing so the author will use the language of the immortal Newton, who, at the close of the Preface to his "Principia", says:

> "I earnestly entreat that all may be read with candour, and that my labours may he examined, not so much with a view to censure, as to supply their defects."

Hughes, Printer,
King's Head Court, Gough Square.

If you have enjoyed this second volume in the
Electric Telegraph Series
you might also like the first.

An Illustrated Handbook to the Electric Telegraph

by

Robert Dodwell

originally published in 1862

www.ingramcontent.com/pod-product-compliance
Lightning Source LLC
Chambersburg PA
CBHW050056230526
45470CB00004B/1555